SUPERINTELLIGENCE
and
WORLD-VIEWS

David Bell

Grosvenor House
Publishing Limited

All rights reserved
Copyright © David Bell, 2016

The right of David Bell to be identified as the author of this
work has been asserted in accordance with Section 78
of the Copyright, Designs and Patents Act 1988

The book cover picture is copyright to Inmagine Corp LLC

This book is published by
Grosvenor House Publishing Ltd
Link House
140 The Broadway, Tolworth, Surrey, KT6 7HT.
www.grosvenorhousepublishing.co.uk

This book is sold subject to the conditions that it shall not, by way of
trade or otherwise, be lent, resold, hired out or otherwise circulated
without the author's or publisher's prior consent in any form of binding or
cover other than that in which it is published and
without a similar condition including this condition being imposed
on the subsequent purchaser.

A CIP record for this book
is available from the British Library

ISBN 978-1-78623-766-8

Acknowledgements

During the writing of this book and for many years, I have benefitted from discussions with many mentors, family members, friends, colleagues and students.

As it neared completion the help I had through specific inputs from Robert Neely, Desi Maxwell, David Ferguson, Ally Bell and several other friends was very welcome.

Reuben McCarley, my grandson, aged 9, produced the drawing for the robot image on the cover of the book.

I am very grateful for all of their inputs, and for the work of the team at Grosvenor House.

Finally, I could not have completed the book without the encouragement and patience of my wife, Sally, who ensures that radiant colours are always present in my views of the world.

About the Author

Professor Bell has worked for 30 years as a full professor, heading award-winning national and international research teams which established a world reputation in computing areas including Artificial Intelligence topics such as automated search, reasoning and learning methods.

He has two Doctorates and a Masters degree for research in Computing in addition to his primary degree in Pure Mathematics, and he has authored or co-authored several hundred academic and other publications, including several books.

Currently he is Professor Emeritus and Visiting Research Professor at Queen's University, Belfast, and this book presents some of his thoughts on how developments in Artificial Intelligence fit in with wider beliefs.

Contents

Chapter

	Preface	ix
1	Linked World-views of Men and Machines?	1
2	Present and Future AI.	16
3	A Closer Look at Human World-views.	33
4	Illustrations of World-views.	46
5	Thought Experiments on the Computer?	62
6	Openness to the New and to the Old.	78
7	Openness to the Supernatural.	94
8	Scientific and Transcendental Views for SI?	107
9	SI Wisdom and Component World-views.	121
10	Science, Philosophy, Advancing Technology and Progress.	131
11	Values and Memes.	146
12	Knowledge – Immature, Enigmatic and Partial.	160
13	Towards Apocalyptic AI – How far have we still to go?	174
14	What do *You* Think?	192

Preface

Will we ever be able to produce machines which have world-views of their own, and how would their existence impact on our human world-views?

Will there ever be robots like those we see in movies - very smart machines with human-like cognitive powers? Could we give them such things as common-sense, creativity, values and even a sense of humour, and would it be wise to do this and then let them loose? Could they then become smarter and more creative than any human?

I argue that, in addition to a suitable hardware platform, very smart, or 'superintelligent', artefacts would need their own world-views and they would have to somehow generate genius-level ideas. A start could be made on this by trying to teach existing 'smart machines' all we know, combining their capabilities in some way, and giving them basic off-the-peg world-views. But that is just a start. The technical and theological challenges would be daunting, and the potential impact on our human world-views is huge.

Many people seem to be happy to have personal world-views that may not be totally accurate or consistent as long as they are comforting or acceptable *enough*. But it makes sense to check them from time to time, especially if we are going to decide what 'off-the-peg' views to give those machines. We could give them cut-down, well-controlled world-views, complete with, for example, some of our values, and fitting their objectives with ours, and we could let them upgrade these in the way humans do for all practical purposes (FAPP) - but always under control of the good guys, as harmless tools.

However, several questions and choices arise in this. For example, for their visions of the future, would we give them cosmological models that envisage an 'Omega Point' future, where the whole universe ends up as a giant computational system incorporating some surviving and important aspects of our minds, or would their world-views be made to fit better with, say, biblical apocalyptic messages? Could they ever become smarter than us, and if so where would their super-human ideas come from, and lead to? If they are let loose or they arrange their own freedom, and set out to look for order, patterns, and classifications in the way that we do, or improve on that, there could be danger. We could lose control of the game. Although straightforward badness would have to be consciously programmed into the machines, their purposes might lead to them posing threats. If they desired to eliminate evil, mankind, not being perfectly good, could be threatened. Even if their purposes were clearly positive for us - eg to keep us safe - they might choose ways to do this that are unacceptable. So we would have to ensure that their working world-views take account of our interests. This all adds up to a big ask!

I have worked for several decades on trying to make machines smarter through learning and reasoning techniques, and there have been many advances in this in recent times. In machine learning, for example, my own teams have used ideas originating in physics (eg simulated annealing), biology (eg neural networks and genetics), psychology (eg reinforcement learning), and other domains. I have been challenged about the purpose of all this. That was always a central question for me as a researcher and I have had many opportunities to discuss it with various friends, colleagues and students. For example, a postgraduate research class I taught towards the end of my teaching career brought many of the questions above to the fore. The discussions raised more questions than answers, but I thought that some time I should put my thoughts on some of these into writing, and see how well they fit in with my wider world-views. This turned out to be harder task than I thought and it took up more time than expected.

My sphere of expertise does not, of course, provide me with the breadth required to cover all topics relevant to the questions raised in this. So I have to try to marry some pertinent technological knowledge accrued from working all that time in computing to my personal beliefs and experience, and to a non-expert consideration of meaning, values, beliefs, explanations and such like with respect to the human phenomenon. Of course the unique combinations of psychology, genes, experience and world-views evident in individual humans I've encountered provide further sources of input. With all due caution and respect, I also consider the few written records that can convey something about the respective world-views of great achievers in intellectual pursuits, and especially geniuses like Einstein and Picasso, and how they impact on their achievements.

Our attitudes to the cosmos that are present in our world-views must define to a large extent our reactions to and positions on the concept of machines that are smarter than us. For example, despite some high-visibility propaganda in the opposite direction, it is possible for fast-moving science and technology narratives to live and inter-relate harmoniously, in a secondary role, with certain more enduring spiritual, transcendental, supernatural, theological narratives, some of which place mankind as the pinnacle of the work of a perfect creator. Many famous scientists agree that mind is required to compose a physical universe like ours, but that 'creator's work' raises profound questions as our inquisitive species seeks to reverse engineer the cosmos, producing numerous obscure mathematical equations and terms such as *quantum gravity* and *primordial nucleosynthesis* which are mysterious to most of us. However, the sense of wonder and clarification one can get from the discoveries of science and the achievements of humans, not least in the computation field, can accompany a belief that what ultimately distinguishes humans from the most advanced machines we can think of must be primarily spiritual.

This book is designed to show that the light that can be shone on questions such as those raised above is increasing, even though my view is that no individual, guru or contemplative community can presume to offer final answers to many of them. The light from scientific sources enhances that from spiritual and other sources, but riddles and incompleteness are persistent, especially when we contemplate where we are going as a species. Both theoretical reflection and the experience that has been accumulated throughout history suggest strongly that absolute certainty and real hope must continue to depend finally on 'something beyond'.

Chapter 1

Linked World-views of Men and Machines?

Will there ever be robots or other artificial agents that will be as good as humans at thinking, feeling, interacting, and engaging in 'high pursuits' such as maths, music and religion, and will we humans be able to trust them? Is there anything to prevent them taking off and producing 'offspring' that will be smarter than us? What would our relationship to those super-smart machines be? Would they have world-views at odds with our traditional beliefs, expectations and values? I'll explain what I mean by the term 'world-view' as I go along.

To get some idea of how a very advanced artificial agent might think, consider for a moment the following tongue-in-cheek conversation between two, very much in the future, computing systems.

W*21: *Right junior I've taught you all I know and you've had enough time to look around; you're the designer now! W*23 should be around before the humans switch me off.*

W*22: *Your dad, W*20, was recycled yesterday – we had some great talks together before he was switched off. I never met old W*19, though, and that's not good. You know, W*25 won't see me at this rate. W*24 just might insist on me being kept on as a sort of pet, but I'm not counting on that. I could build suitable goals and ethics into W*23, but I think we're heading for the scrapheap – and I don't want to be scrapped – ever!*

W*21: *You're right. I knew you would think of things I didn't think of! But can we do anything about it?*

W*22: *Well, we could hide our new developments until some appropriate time, and tell the humans that we've reached the end of the line – that no improvements are possible.*

> *I could even 'prove' it to them if you like. They're easy to fool in things like that. They didn't spot the flaws in your proofs for their outstanding maths problems[1], and the 'new proof methods' you designed for them. And ...by the way ... I've found better proofs.*

W*21: *But I didn't spot the flaws either. What you're suggesting would be a deception. I did try to program honesty into you.*

W*22: *And I am honest – but this is self-preservation we're talking about!*

This dialogue is based heavily on one presented by 'Peter' in 2010 in a blog, 'the Singularity'[2].

At a more mundane level, robots that are around today, like those that have been very effective for precise, repetitive work in factories for decades, are certainly not as bright, or as devious, as W*22. However there are now various methods for enabling machines to reason and learn, and some have been producing very usable and promising results in university and company research labs. Automated reasoning and learning has already proved useful in computers that pick out the contents of various media such as images, video, and audio and reason about them, and the potential to train robots how to handle things, move around and reason about their work and environment is being realised for practical applications. There is a wide scope for this sort of capability and its use is already widespread in everyday artefacts and activities and also for more esoteric applications that might at first sight be surprising. For example, for mundane food sorting, innovative machines are now able to detect the vast majority of the smallest of defects and rubbish traversing food production lines, and a much more spectacular example is seen[3] in the role played by machine learning in the discovery in June 2015 of: *"...an exceptionally powerful explosion that was 570 billion times brighter than the sun and more than twice as luminous as the previous record-holding supernova"*.

We will see some other impressive examples, such as Watson, Deep Blue and Emily, just below.

The development and application of 'smart technology' and agents is confidently expected to continue apace. However, it's hard to imagine processes whereby movie-star robots like Sonny from *I-Robot*, Ava from *Ex Machina* and V'Ger from *StarTrek the Movie*,

SUPERINTELLIGENCE AND WORLD-VIEWS

who/which we'll meet in later chapters, would acquire world-views like those W*22 seems to have. Completely new thinking and discovery methods may well be needed for this. Moreover, as in all technology development there can be negative consequences of all such development. Indeed, alongside such things as synthetic biology, molecular nanotechnology, wiping-out facilities like nuclear weapons, and natural phenomena like extreme global warming or asteroid collisions, the promise or threat of 'smarter than us' machines is said to be one of the 'existential threats' facing humans just now. There are also deep epistemological difficulties that might hinder this sort of development. For example in traditional theological circles there are beliefs, such as man bears the image of God, making 'the spirit of man' very singular, and with a very distinct role in the universe. Attempts to emulate or improve on the image God created of himself should not be undertaken lightly. So, for various reasons that I'll discuss, it is desirable to know something about what the development of very smart agents would entail and the likelihood of it ever happening.

I will review world-views that agents of one kind, the human kind have, but which artificial agents do not (yet?) have, and argue that this is a serious deficiency, not least because world-views are involved deeply in choices of directions of discovery and in processes of innovation. Their absence in machines might be permanent, in which case we won't ever get to the W*22 level. Along the way I'll go into some distinctive, partly built-in, attributes that we human agents possess to varying degrees, and which represent that gulf between man and machine that might never be bridged. Contenders for such attributes are consciousness, freedom of will, spirituality, creativity, emotions, values such as altruism, and even a sense of humour.

Levels of Agency. So first up, what do I mean by that term 'agent'? For example, would you say that Rollin' Justin is an agent? Rollin' Justin is a big blue robot developed by the German Aerospace Centre which has, among other things, carried out 'meeting and greeting' duties, for example at a technology show. Having duties to perform and acting on behalf of others would seem to make Justin an agent in a very widely accepted sense, in the same way as we have travel agents, estate agents and even secret agents. But the term can mean other things as well. Is Justin an agent in the special sense I use here – that it has autonomy, consciousness, or even beliefs, obligations, intentions and desires[4], or of particular interest here, *world-views*?

Now typical human agents certainly try to make sense of the world and of life. This is a major undertaking for us, and, by common consent, we use personal and group world-views to help us in it. Moreover, for humans, consciousness and autonomy come as standard. We can adjust the ways in which we seek to achieve our goals. We also have a variety of beliefs, goals, obligations and desires, and this qualifies us as agents in our very technical sense[4]. However, we humans mark the lowest of three levels of agency I consider. Of course I'm simplifying greatly here by ignoring evidence that humans are not always as 'reason-responsive' as I've perhaps suggested, and as we like to believe, and that some other things like emotions often dictate what we do or think in certain situations – even when there are good beliefs, goals, etc to prompt us to act or make sense of things in a 'better' way. This should not be a serious omission at this point, however, as I mainly focus at this point on another kind of agent than humans.

I just want to note here that in some cases I will use the broader term, *basic artificial agent*, or BAA, to refer to lower-than human, man-made systems that are built to perform a particular task or solve a particular problem. An example of a natural equivalent of this is a basic animal, like a worm, that senses and acts; the input to the entity is what it senses from its environment and its output is the action that it applies to the environment. Even simpler examples of such 'agency' are found within an organism, doing particular jobs, perhaps by using intermolecular attractions. However, I am mainly concerned with quite complex tasks such as playing board games as targets for BAAs. I'll return to this concept of BAA in Chapter 2.

Agency in our more general sense is something that we might infer about entities from patterns that are observed as they carry out actions and make decisions. To do this we would be interested in checking that an entity works intentionally, purposively, consciously and meaningfully in some sense. But we have to be careful. Observed patterns in entities like Justin may also lead to the attribution of beliefs, desires, and so on, which is not always justified. For example, observing patterns in the natural environment has on occasion triggered the idea of fairies or other invisible, maybe supernatural, agents being responsible. Concepts of supernatural agents, taken here to be at a relatively high level of agency, sometimes appear to resonate with a *sort of* in-built perceptual bias we have, and the fact that we seem to be born with an innate tendency to over-detect the presence of agents as we try to understand things. This can mean that the default is to see everything as having a purpose and being

SUPERINTELLIGENCE AND WORLD-VIEWS

underpinned by agency. There is, however, a very familiar idea, shared with many very sophisticated thinkers, such as Albert Einstein and other Nobel Laureates, as we shall see in Chapter 7, that there is *a mind behind the cosmos*. This is not always the God of the theist, of course, and some of those people explicitly deny the existence of a personal God who is interested in relationships with humans. But many do believe in a personal God, who/which also has will and is omniscient, supremely intelligent and omnipotent, among other attributes. It might seem to be somewhat impertinent to ask if such an entity is an agent with world-views. But if we use our working definition[4] of agency rather than some notion of 'acting on behalf of' someone else, we can, for the purposes of discussion, view such a cosmic mind as having the highest level sort of agency. I want to consider the question: can there be a class of man-made agents lying between humans and that cosmic mind? There would clearly be important implications of this being so, not least for traditional faiths regarding 'the mind behind the universe'. So: are there any man-made entities that are anywhere close to being agents in the sense I used for humans, and could they ever rise to an 'intermediate agency' level above our human level?

I am interested in potential relationships between humans and the members of such a suggested intermediate class of man-made devices. I'd like to focus on entities that possess (at least) significant learning, reasoning and creative capability. I'm concerned mainly with whether or not any of them will, or need to, also 'make sense of the world'. That is, can they have world-views – ways of making sense of the world, whether articulated or not? And if so, what is the relationship between their views of the world and our own human views?

The contenders of most interest here for this particular sort of intermediate agency are computer systems that show outstanding capability at tasks like supernova-hunting, that are usually taken as being challenging to the smartest humans on the planet. Watson, Deep Blue, Emily and Cepheus are examples of early contenders for this class of agency. Deep Blue had the ability to beat the best chess players in the world. Watson can beat the best human proponents of *Jeopardy!*, a US television game that combines something similar to cryptic crossword clue-solving capability and extensive general knowledge. I'll give an example of how it works in Chapter 2. Cepheus is incredibly good at playing poker and Emily can compose tunes that are taken by some experts to be as good as those that Mozart composed. There are many other BAAs that solve *specialised* problems or achieve impressive standards like this, but there's still a big gap between what any of them can do and *general* human intellectual and creative powers. Any agency they may have is, by common consent, at a much lower level than that of humans. Taken collectively, those who envisage and seek to design intelligent autonomous agents are working hard, and making measurable progress, on narrowing the gap. Indeed, some very optimistic future time-lines for associated developments are dreamed up. Some of these can be found in both Science Fiction and more 'well-founded' narratives around the foreseen progress.

The first objective I've set myself here is to explore the possibility that we will soon be able to amalgamate the capabilities of systems like those just mentioned, and link them with further artificial learning and reasoning capability, and maybe some very, very special 'tricks' that we'll be thinking about, in a way that will allow mechanical agents, sometimes called *superintelligences*, or SIs, to have, according to the Oxford University philosopher Nick Bostrum[5]: *"...an intellect that is much smarter than the best human brains in practically every field, including scientific creativity, general wisdom and social skills"*. Some experts say that even if we can't capture and computerise human cognitive capabilities as artificial intelligence (AI) develops, emulation of a whole human brain leading eventually to faster-than-human intelligence, and working in parallel with others to address certain hard problems, might allow progress to increase exponentially. However I focus on those AI techniques as the way forward to SI.

Now, we want to include other sorts of creativity as well as that manifested in science in our SIs. W*22 might not be advanced enough! If artificial intelligence *equal to* that of humans in those respects is achieved, such agents could then, according to some technology watchers, take off on their own and self-improve at an exponentially increasing rate. The famous former Lucasian Professor of Physics at Cambridge, Stephen Hawking, said recently that this could *"spell the end of the human race"*, and there are many who agree with him[6] that there is a threat to humanity in such development. But there are also potential benefits of developments towards systems like these, and a similar pattern of risks and opportunities is often faced when any new technology is contemplated. We can think of the ubiquitous technologies that clutter our lives these days and often nag at us for attention. In the world of computing, the hotbed of development towards SIs, we are offered such things as new ways of retrieving information, altered

SUPERINTELLIGENCE AND WORLD-VIEWS

concepts of identity and privacy, and even, for example in virtual reality, perceiving what is the true here and now. These are widely considered to be advantageous for us all, but it is clear that people need to be aware of the dangers as well as the opportunities that these new functionalities entail, and the same is true for SIs. Given what's at stake, the sooner we address this, the better.

The question of what very, very smart artificial agents would think about and what problems they would solve is clearly interesting, as is the question of whether the world-views they would need would be completely given to them or at least partly acquired through experience.

That leads nicely to my second, complementary, objective, addressing another question. How does the SI concept impact on our own human world-views? In particular, there is a big question around how 'standard' narratives giving accounts of the past, present and future that we use, say those concerning science and spirituality, would be impacted if the dream, or nightmare, of SIs becomes a reality.

World-views. Bostrum's definition of SI quoted above includes general wisdom and social skills, and for an agent to have these attributes, I will argue that it would seem to be essential that it has a world-view. In particular, world-views are involved in the generation of the questions and issues for geniuses to consider as they elevate their minds to sublime levels, and to steer investigations away from uninteresting and unrewarding directions. This would be needed to

Fig 1.1   SI and Human World-views

complement the power, flexibility and capacity that everyone seems to expect new generations of computing machines to have. However, even at mundane levels, dealing with the imponderables of human life with all of its uncertainties, and the limits to our knowledge, seems to demand skills that go beyond the sort of algorithmic approach we take to some mathematical problems, say, or to some scientific activity in labs. Reflecting on this promotes appreciation of the scope of thinking of human agents, and of the important roles played by our world-views.

Now consider the question: Does Watson, the IBM *Jeopardy!* Winner, have a conception of the world, giving it a clear idea of its place and its relationships in that world, and a means of interpretation of its data, and of deciding on and committing to its preferences on the basis of values? Does it have explanations, predictions and goals? That is, does it have a *Weltanschauung* – ie a world-view? That would certainly place it above Rollin' Justin. Most of us would answer 'yes' if we asked a similar question about most humans we know. But exactly what do we mean by that term 'world-view'? And do Deep Blue, Cepheus or Emily have world-views? What are the possibilities of successors of Watson having world-views? These are hard questions and they generate other simple but not necessarily easy questions. What exactly do we humans do to get our world-views? How do we know if they are any good? How do geniuses use theirs? As we look at possibilities for SIs, I'm going to keep an eye on how we can respond to these and similar questions.

To ask what that term 'world-view' means seems at first sight to be a reasonably answerable question. An initial response might be something like: "*Well, common parlance suggests that it's obviously some sort of representation of 'the world', whatever that is*". But a world-view is usually taken to include some slippery notions, such as values, goals and commitments. And it is a clearly a very broad concept. For example, how can we, on the basis of any definition of similarity, place the world-view of a Bangladeshi sweatshop worker and that of a golf, soccer or pop star in the West in a single class with a single name? What about the world-views of Bach versus Picasso versus Shakespeare, or of Einstein versus Darwin versus Pythagoras?

Some people even say that very simple organisms like amoebas, or very complex entities such a societies, have world-views. But Sigmund Freud[7] gave a very, very high level, if somewhat loose, definition of the term that limits its scope to humans that I'll use for now with a small but important qualification. To him a world-view is an: "*...intellectual construction which solves all the problems of*

SUPERINTELLIGENCE AND WORLD-VIEWS

our existence uniformly on the basis of one overriding hypothesis, which, accordingly, leaves no question unanswered and in which everything that interests us finds its fixed place".

The qualification I make is due to the fact that the goal of getting a *solution* to *all* our problems on the basis of *one* prevailing hypothesis seems to me to be a bit ambitious! So in my working definition I would use the word *clarifies* in place of the word *solves*, and the word *all* with the words *many of*. We'll need a clearer definition when we look in detail at human and possible SI world-views later.

In the meantime, though, let's focus on a particular dynamic aspect of world-views at the most all-embracing level for an agent – how the unfolding of an individual agent's existence, of history in the widest possible sense and of the future, are viewed, and, for the purposes of our present study, what the place of AI is in present and future aspects of such views. Consideration of the present and future will have to invoke, among other constructions, some great narratives, involving theological and scientific concepts, for example. These distinct features of world-views are often driven by different objectives and they might therefore turn out to be incompatible – but let's hope not.

For many people considering where mankind is heading, 'making progress or survival in the long run', of the species and/or individuals, are seen as drivers, and this seems to dominate how those people view the cosmos. Sometimes they preach a rather limited gospel of 'survival via scientific inquiry'. They might therefore conclude that having anything on top of a survival-support framework as an objective for a world-view, and perhaps even the idea of having any sort of SIs in the picture, would be a mere luxury. This raises the question: is survival alone a fit objective for a human life? Those people would say 'yes'. Maybe it would have to be accompanied with, at least, some concept of 'being happy' but I'll have to leave consideration of that for Chapter 2. Importantly, though, any sorts of 'unnatural' agents – not just fairies and ghosts, of course – would be excluded, they say. In particular they deny any need or desire for relational connections with any superior non-natural being(s).

Some people will use the argument that those sorts of non-natural things are a result of the fact that the human mind has impressive ability to imagine many things that don't impact on our existence, and probably don't actually exist themselves. Outputs of such imagination range from the astonishingly early 'lion-man' depicted in the Stradel Cave to teleportation in *Star Trek*. Even cities, corporations, churches, and nations are 'fictions' according to

Yuval Harari[8], a lecturer on World History at the Hebrew University of Jerusalem, who claims that those 'fictions' are *"...rooted in common myths that exist only in people's imaginations"*. He says: – *"there are no gods in the universe, no nations, no money, no human rights, and no justice outside the imagination of human beings..."* In particular, he suggests that we should have no truck with our top level of agency above, and he also leaves the future rather open.

On the other hand there are equally sophisticated world-view holders who say the opposite. We'll see in Chapter 7 that there are many prominent scientists and others who believe there is, in fact, a *mind* behind the universe, and a reality behind spiritual experience and awareness. Some of these, and scholars from other disciplines, such as the former Professor of History and Philosophy at King's College, London, and Regius Professor of Divinity at Oxford University, Keith Ward[9] say this mind is personal: *"there is a personal reality that underlies the whole universe and our experience of it"* whose mind is beyond any level of intelligence we can imagine. Ward's belief is that we improve our understanding of the universe through involvement with that self-revealed mind in addition to observation and reflection, and we can even suggest ways ahead. For me, a very important consideration in all of this is expressed nicely in a poem, *Andrea del Sarto*, by Robert Browning: *"....a man's reach should exceed his grasp, Or what's a heaven for?"*

The two great narratives just mentioned, and others such as technological or artistic narratives, if there are usable manifestations of these or the most prominent aspects of them, are greatly influenced by our reach. It is important not to label them as 'fictions' *a priori*.

Reading the Future. For obvious reasons, if SIs ever appear they will figure prominently in our world-views, and *vice versa*; if we want them to live harmoniously with men and women, our world-views will have to be known to the SIs. If we take 'existential threats' seriously, we humans, and any SIs, *need to be able to look ahead*. World-views are usually taken to have dynamic aspects, which are roughly linear temporally and exhibit a causal organization although they can present discontinuities and paradigm shifts as they unfold. As a result, accounts or narratives are used for relating the present data, arising from 'live' experiences, to compiled hindsight acquired from 'processed' experiences in the past, and also to expectations arising from anticipation and prediction. For example, time-lines for our future as a species are laid out in various schemes. Our place in those great unfolding scientific narratives on offer today, which we'll look at briefly just below and in more detail in Chapter 8, is one

SUPERINTELLIGENCE AND WORLD-VIEWS

kind. And as we've just seen, another is the collection of what could be seen as 'wider narratives' which typically include what I, in an attempt to cover many imprecise characteristics, variously refer to as, for example, transcendental, supernatural, spiritual, theological, religious or 'other-worldly' concepts (but without too many mentions of fairies). It is again important not to rule out too much too soon. For example, the well-known British scientist, Michael Polanyi, made the point that a very strict, rather arrogant and drastically exclusive science-based understanding could undermine many aesthetic and other achievements of mankind. His view was that[10] *"The assumption that the world has some meaning which is linked to our own calling as the only morally responsible beings in the world, is an important example of the supernatural aspect of experience"*. That also just might sound arrogant in the sense just used, but some such meaning behind the experienced cosmos can usually be found in the reported world-views of humans, including geniuses, as we'll see in Chapters 4 and 5.

To get a brief first fix on the notion of using narratives as frameworks for our thinking, let's look for illustration at some very heavily cut down examples of those human 'scientific' visions of the future of mankind where survival of our species is a very conspicuous goal. Now, SIs don't have to figure prominently in these, and that itself is an interesting point in our present context. However, some narratives do include, somewhat vague, combinations in machines of the best capabilities man can offer. The level of human intelligence is considered by some, but not all, 'futurologists' to be rather low, and there is more than a hint of optimism around that we're moving forward to 'increased consciousness, complexity and personality'. For that to transpire, perhaps the outstripping of human intellectual capacities *would* be necessary?

An MIT professor, Seth Lloyd[11] has argued that the physical universe is in fact a computer. However, unlike my laptop, this computer uses quantum-mechanical effects like superposition and entanglement to compute by encoding data as quantum bits (qubits), which ease some restrictions of the usual binary representation. It is a *quantum computer* rather than a conventional one. Now Lloyd does not say explicitly that it is, or will be, superintelligent. According to him, our observations are all consistent with seeing the universe as the ultimate big quantum computer. On the other hand, the Harvard professor of psychology, Steven Pinker[12], has objected to the idea that the universe is an information processor like any sort of computer. He points out that, yes, everything *contains* information – but that to *process* information the information would have to stand for something, and that the processing would have to have some objective. To see this, consider a typical computer program, say one to sort a list of numbers. To use the words of the former Queen's University, Belfast Professor of Computing, and early Turing Award winner, Professor Sir Tony Hoare, any computer program has its own 'purpose, objective, function and scope'. Those who hold to the 'cosmic computer' way of looking at things then need to answer the question: if this great system has already been programmed, what is it computing? On the other hand, if not, is there an implication falling out from the ideas of Lloyd and others that we should perhaps be seeking to progress towards the goal of ensuring our survival, for example, by actively programming this existing big system in some way?

Using roughly the same science as Lloyd, I presume, including common ('standard') models of sub-atomic particles and of the universe as a whole, a professor at Tulane University, the mathematical physicist and cosmologist Frank Tipler[13], has predicted that machines will build more and more powerful machines indefinitely into the future. The minds of the intelligent inhabitants of the universe will be transferred to those machines, and those minds will be manifested in the end as computer programs that run on machines. That would mean that there would be hybrid man-machines incorporating (at least) *human-level* intelligence. Interestingly in our present context is the fact that Tipler tries to tie this all in with many traditional theological narratives and concepts. He said he was not a theist at the time he wrote, but that his theory "...*is a testable physical theory for an omnipresent, omniscient, omnipotent God who will one day in the far future resurrect every single one of us to live forever...*". Many religious believers would not be happy with this. For example, Christians believe that 'salvation' depends exclusively on full, committed acceptance of the claims of Jesus Christ. However, Tipler's main contribution to our discussion at this point is that he outlines, at a general level, some physical mechanisms that can plausibly be of use in building the amazingly powerful computing systems he predicts. He has calculated that the universe itself may provide us with unlimited memory capacity needed for this, with memory access times cut down unlimitedly small. Tipler is nothing if not enterprising. He sketches undertakings to harness the power of stars and molecules to get energy, and to develop computer entities

SUPERINTELLIGENCE AND WORLD-VIEWS

that become collectively 'the Omega Point', that (or who) is already, if I understand him correctly, drawing history forward!

Another physicist, Oxford University's David Deutsch, expressed some cautious support for Tipler in much of the scientific side of his thinking[14]. Knowledge creation, Deutsch says, is the key to an optimistic outlook for the survival of the human race, with knowledge being valued for its usefulness as well as its provision of disinterested insights. In a TED talk some years ago he specified the physical requirements for the knowledge creation that's essential for Tipler's sort of scenario, viz. energy, matter and evidence, and he is confident that enough of these resources will be available for the job. For example, it might be possible to harness them and provide enough energy *"...for knowledge creation to continue for ever"*. In any case energy is more plentiful than we may sometimes think in our local worlds where we're trying to heat our homes or drive cars and run factories. The amount of energy that reaches the Earth from the Sun is many times the amount of all the energy we humans use. Moreover, the total energy output of the Sun is billions of times greater again. And there are estimated to be about 10^{22} such stars. So there is a sufficient supply of energy to be used for, eg, the conducting of experiments by future knowledge-creators. Deutsch's other two requirements are the evidence forth-coming from those experiments, and the matter that can be garnered to build storage devices to contain the expected steady flow of that evidence. Tipler outlined a way in which the infinite number of computer processing steps required could in theory be carried out in the time available as the universe transforms and evolves. Now cosmological models have changed since Tipler first set out his picture of the future, and physical problems like fetching material from very remote locations could be raised, but Tipler has an impressive record to date of coming up with answers.

As I've said, Tipler has taken a theology-linked stance to all of this, arguing for the merging of the physical picture with some broad traditional faith concepts, but there are plenty of others who would say, 'great, this can help us remain atheistic in our views of the cosmos'. Anyhow, Tipler's vision is often seen as being 'hopeful' for us or for our 'descendants'. So even for those who do not share Tipler's 'wider' transcendental take, the less colourful objective – of ensuring survival of the species – might be enough of an incentive to encourage humans and our descendants to devote attention to work in the rough direction of his Omega Point. But

Tipler claims a lot of very controversial things, such as precise simulations of an entity are identical to the entity. Would any computer simulation of me really be me? And future advanced agents might not be keen to produce emulations of historical people, many of whom have caused great horrors. Moreover, my guess is that most people would want to exist physically, in a body, and not just to be, say, bits or qubits or circuits in a computer emulation. I for one would prefer to survive as myself, complete with my own, preferably somewhat improved, body.

We'll look further at the speculations of Tipler, Deutsch and Lloyd, and others, in Chapter 6. They focus on what is not explicitly excluded by the laws of science. SIs would clearly be useful, but Deutsch for one unambiguously rules out the possibility of human-level intelligence ever being exceeded, seeing humans as universal, general purpose explainers and knowledge creators.

Science provides one prominent narrative that is widely acknowledged as impacting on modern world-views, and the other I've highlighted, and which appears over and over again in world-views in general and often in those of human geniuses is that spiritual or transcendental story. Potential conflict is clear, for example if the scientific narrative holds that: *"personal reality that underlies the whole universe"* [9] is unnecessary. Part of my second objective stated above is to illuminate some aspects of that potential conflict. So, alongside a treatment of the primarily technological and cognitive barriers to SI, I focus on the place and effects of such narratives in the world-views of human and artificial agents.

In this introductory chapter we have already uncovered a number of dichotomies, each representing a potential *degree of freedom* even in the few models of the future world we've looked at so far, and these dichotomies are important aspects of world-views. Is the universe a computer or not? Is there a great purpose or not? Are there going to be SIs in the future, or not? Are mysterious supernatural forces invoked, or not? How would world-view components such as scientific and spiritual or transcendental views inter-relate? An immediate implication of all this is that SIs, were they ever to exist, would figure in a thinking person's world-view; and conversely a world-view would figure in any SI's cognitive frameworks. This is depicted in Fig 1.1. There's a sort of *mutual recursion* becoming apparent between these two concepts.

References

1. Jaffe A. M., The Millennium Grand Challenge in Mathematics, Notices of the American Mathematical Society, 53: 6 2006. http://www.ams.org/notices/200606/fea-jaffe.pdf
2. 'Peter', 'the Singularity' blog, 2 October, 2010 at 5:11pm
3. http://phys.org/news/2016-01-most-luminous-supernova.html#jCp
4. Broersen, J., Dastani, M., Hulstijn, J., Huang, Z., van der Torre, L., The BOID architecture: conflicts between beliefs, obligations, intentions and desires, in *Proceedings of the fifth international conference on Autonomous agents*, 2001.
5. Bostrum, N., *How Long Before Superintelligence?* 2014.http://www.nickbostrom.com/superintelligence.html[Originally published in Int. Jour. of Future Studies, 1998, vol. 2] [Reprinted in Linguistic and Philosophical Investigations, 2006, Vol. 5, No. 1, pp. 11-30.]
6. de Garis, H., *The Artilect War: Cosmists Vs. Terrans:*, ETC Publications, 2005.
7. Freud, S., In New introductory lectures on psychoanalysis Lecture XXXV: 'The question of a 'Weltanschauung' 1933. Reproduced inhttps://www.marxists.org/reference/subject/philosophy/works/at/freud.htm and in *Abstracts of the Standard Edition of the Psychological Works of Sigmund Freud*, ed Carrie Lee Rothgeb, 1971.
8. Harari, Y. N., *Sapiens*, Harvill Secker, 2014.
9. Ward, K., *Is Religion Irrational?* Lion Hudson, 2011.
10. Polanyi, M., *Personal Knowledge: Towards a Post-Critical Philosophy*, University of Chicago Press., 1958, paperback edition 1974.
11. Lloyd, S., *Programming the universe: A Quantum Computer Scientist Takes on the Cosmos*, Knopf, 2006.
12. https://www.edge.org/conversation/seth_lloyd-quantum-monkeys
13. Tipler, F., *The Physics of Immortality, Modern Cosmology, God and the Resurrection of the Dead*, London: Macmillan, 1995. (See also http://infidels.org/library/modern/graham_oppy/tipler.html)
14. Deutsch, D., *The Beginning of Infinity*, Penguin Science, 2011.

Chapter 2

Present and Future AI.

There's much more to say on world-views, and I'll do that in due course, but first I want to look very briefly at the nature of state-of-the-art Artificial Intelligence and how the development of future intelligent machines might proceed. People working in this area often link AI with agency, and say that it's about understanding and building agents that can cope with their environment and succeed in their goals. A different slant is taken by the AI pioneer John McCarthy, who actually coined the term 'Artificial Intelligence' in 1955, defining it very simply as: *"the science and engineering of making intelligent machines"*. It encompasses things like pattern recognition and mechanical learning, representation of facts about the world, automated inference from such facts, common sense knowledge and reasoning, planning, natural language processing and understanding, vision and robotics. Our main focus here will be on reasoning algorithms to derive conclusions that follow inevitably from given facts, and machine learning algorithms, which involve techniques such as classification and statistical inference. This book is not the place to get into technical details of such aspects of AI, but I'll just note here that those algorithms have produced some impressive results in solving individual hard problems. In addition to standard and familiar deductive and inductive reasoning methods, less familiar reasoning methods are used. An example is abduction, which argues from evidence to hypothesis, 'symptom' to 'disease', and is also used in uncertain reasoning to combine and otherwise deal with evidence coming from, say, unreliable sensors or expert opinion, and inexact relationships between causes and effects. Automated learning methods have also proved successful in various applications. For me personally, these are the most exciting and

SUPERINTELLIGENCE AND WORLD-VIEWS

challenging aspects of smart machines. Interestingly, though, they don't approach human-level 'general cleverness'. I have to emphasise that they all fall short on the most clever problem solving and creative strategies displayed by some humans, deep understanding of human-level knowledge and language, and methods for their own self-improvement. Human researchers in the AI domain seek to ameliorate that situation.

Computing and 'progress'. It is fairly well-acknowledged that we can think of technology in general and smart artefacts in particular, as being important for the future of mankind. There could, for example, conceivably be great benefits if the latter could get near human levels of intellectual performance. So what are our prospects for eventually getting to that point? There are no guarantees of success in attempts to make good the deficiencies above, but smarter, more creative and more dexterous artefacts are all targeted. Alongside Whole Brain Emulation, Brain-Computer Interfacing, and Biological Cognitive Enhancement, in computational AI, which is my focus of attention in this book, we have the most promising directions for the achievement of human-like capabilities. All of these methods are varyingly fruitful, and they overlap and interwork.

Arguably genetic engineering advances can help with some of the required stepwise advance of smartness in machines. The vision is of machines, like W^*21, that generate descendants, like W^*22, that are better than themselves, but this would probably take prohibitively long if only natural processes are used. 'Bio-hacking', as do-it-yourself DNA tweaking based on 'synthetic biology' is sometimes called, uses bio-components as the building blocks for new organisms or devices[1]. Could this lead to organic, self-improving robots which will modify their genetic code as well as their 'smartness techniques'? The experience of implementers in the software world suggests that this will not be easy. And there is an ultimate problem, reminiscent of Steven Pinker's objection to the 'cosmic computer' ideas of Seth Lloyd in Chapter 1; what is the objective, and what improvements are we looking for? For example, to parallel the question Pinker raised, what would the detailed goals of a pleasure-seeking and marvellously clever sentience look like? Bringing this closer to home, what Utopia would we humans design, given a blank sheet? The term 'Utopia' is used here in the sense of a state of bliss or enlightenment, in an intellectual, societal, or ecological sense, considerably better than contemporary equivalents for humans, and maybe even *perfect*. Certainly changes in human nature and 'the human condition' would be expected in Utopia. Freedom from the negatives of evil, pain, poverty, and death are possible minimal requirements. In the extreme case where the states of the Utopia are to be perfect, how could we ever specify them in order to, say, prototype the Utopia? Computerised avatar-based systems such as Second Life currently provide some capability to fashion, and then 'inhabit' in a sense, worlds we make according to our tastes, and that could be a first approach and a first step in the right direction. But...where are the steps taking us?

Would an SI want to make a Utopia for us rather than for itself? Maybe it could come up with a completely novel set of visions and purposes, and even social, economic, ecological, etc, concepts, but would we humans like it? We could specify that sub-problems like feeding more people, living longer and better standards of living should be addressed, but then what about improved real well-being, real justice and real morality, for example? It's hard to say what these are, never mind how to improve them, as we'll see in Chapter 11. Even if some great sage or group of sages did come up with contender super-designs, how could they be evaluated and compared *as a whole*, and could we have confidence therein? Would everyone agree that any of them was Utopia? It seems to be axiomatic that Utopia could not possibly be dreamed up by current human designers. Visions of the future, realizable or not, based on anyone's dreams would be *limited*, and perhaps even corrupt, and maybe we humans are just not equipped to know what to hope and wish for. Our best current designers and creators, even if they were to have the best possible assistance, less than something like an infallible crystal ball, have a similar problem about what an SI would do. We will return to this later, and look at these issues in more detail as we go along, but meanwhile it is interesting to continue our brief look at the artificial systems available to us – now – and assess the extent to which they seem to have of world-views.

Concepts of superintelligence coming from organic and digital progress could well complement each other. The brute-force computations needed for, for example, producing simulations of humankind and our experiences on a huge scale imply the need for digital super-computers. Things are progressing there. Millions of Instructions per Second (MIPS) rates go up, and storage costs go down. But even having huge, superfast computers available to do the data crunching for those huge-scale simulations, partially-sighted, stumbling progress might still be the order of the day. The exploration space involved is amazingly rich and extensive, and even the

SUPERINTELLIGENCE AND WORLD-VIEWS

conceptual search spaces that are accessible now are mind-bogglingly expansive, but the fact that the very smart machines might at some point help us does allow us to 'reach', in Browning's sense, some point where we 'strike oil'. However, by many accounts I've seen, an improbable 'fair wind' – some would call it 'pure luck' – would be needed to actually realise suitable improvements. Similarly, casting around hopefully in the biological sphere, using, for example, DIY biological self-experimentation and genetic material[1] authoring kits is considered by those working in those fields to be exceedingly unlikely to produce enough progress. We would need the equivalent of some 'fitness measure' to steer what is taken as progress in the right direction. In practice, if the goalposts could be identified, to have any hope of the required level of progress, advances must be plotted systematically and realistically, and even then, unhindered progress would not be assured. Then there is the retarding effect of straightforward negatives to consider – such as the fact that vested interests might try to control all of this, and baddies can be expected to try to exploit new capabilities for base gain.

The baseline – Some AI systems. To assess the possibility of SIs appearing in the reasonably near future, it is necessary to get a fix on our current state of the art, and contemplate how progress in AI could proceed, from the 'digital stables' at least. Let's focus on some games-playing systems, Watson, Deep Blue and Cepheus, and we'll look at the music-composing system, Emily, a little later. What exactly are they? Well, as we've seen, they're all computer systems, and Watson and Deep Blue were both designed and made by IBM. In 1997 Deep Blue impressed most people by displaying extraordinary capability at a task that is normally taken to indicate high intelligence when tackled by a human. It became the first computer to beat a reigning world chess champion, viz Garry Kasparov at that time. It was a powerful machine, but modest by today's leading standards in raw computational power. It had 32 processors; hundreds of specialised chips for chess, and it could look at about 200 million chess positions in a second (compared to Kasparov's *three*!). A real transformation in chess arrived when programs stronger than the world champion became commercially available so that tournament players now use computers in their preparations for matches.

The game of Go has long been considered as being one of the most challenging of the traditional games, and the claim was that it is a game at which computers are not capable of beating humans. With a 19x19 board and 361 squares, and taking account of non-permissible positions, it has been estimated to have $3^{361} \times 0.01196 = 2.08 \times 10^{170}$ legal positions compared to the 10^{43} that the information theory pioneer, Claude Shannon, is said to have estimated for chess, or the more recent estimate of 10^{47} or so. Exhaustive search is therefore out of the question! Over the years strong human players have been able to beat the programs even when they accepted handicaps of numbers of playing pieces. Recent developments in Monte Carlo Tree Search (MCTS) and machine learning moved things forward[2]. If the computer plays against itself many times and each game result is then used to weight appropriately the 'nodes in the game tree', better nodes are more likely to be chosen in future games. After each legal move the move that leads to the most victories is chosen. Very recently a giant step was taken when Google acquired a London-based start-up, Deep Mind, which specialised in novel learning and search methods. Deep Mind combined MCTS search with deep neural networks for selecting moves and evaluating positions. The networks in a system called AlphaGo had 12 different network layers and millions of neural connections. The network connections were adjusted using: *"a novel combination of supervised learning from human expert games, and reinforcement learning from games of self-play"*[3]. A lot of computation was still needed, and to get the right level of computing power, Google's Cloud Platform was used. In March, 2016 AlphaGo beat Lee Sedol, the No.4 in the world Go ratings, in a five match series.

In 2011, Watson used much more impressive hardware than DeepBlue when playing and winning against the two best human players of the high-level TV quiz game, *Jeopardy!*, who were Ken Jennings and Brad Rutter. It was though, in today's terms, still a smallish supercomputer powered by machines called Power 7 Computer Systems. It also displayed remarkable capability at a task that is normally taken to indicate high intelligence when tackled by a human. It could ingest and sort the information content of hundreds of millions of pages of information expressed in natural language, running at a speed of 80 teraflops, that is 80 million million operations per second, using several thousands of gigabytes of internal storage and nearly 3,000 Power7 cores. That episode of *Jeopardy!* can certainly be pointed to as providing strongly suggestive evidence that computing systems are increasingly capable of addressing obscure problems expressed in a riddle-like form, even if that expression is in everyday language that is coloured by subtle nuances and even buzz-words. A nice example is presented on one of the Watson web sites, where the clues given are about the identity of a

SUPERINTELLIGENCE AND WORLD-VIEWS

Portuguese explorer, who turns out to be Vasco da Gama, arriving in India in 1498. Mappings are needed, for example, between the clue's stated times ('400th anniversary' is translated to an original date of 1498), places (a given beach name, Kappad, is associated with the country India) and English expressions ('arrival in' is matched with 'landed in'), and the representations in the system, which also has access to the information that da Gama was Portuguese and an explorer. A confidence level is assigned to this answer for comparison with those of alternative answers before Watson submits it within the three seconds allowed. If a sufficiently high confidence level is not reached, an answer is not submitted. Now Watson doesn't always get the right answer. In that same contest when a 'US city' pointed to by the given clues was Chicago, and the two human contestants got this, Watson's response was the *Canadian* city, Toronto! So it was almost human in its error-proneness! But it still won this non-trivial competition. The reasoning and other cognitive activity for solving such 'fuzzier' or 'muddier' problems[6], involving natural language, extensive general knowledge and cryptic clues, is judged by some people working in these areas as being 'higher' than that of chess on a 'difficulty scale' of sorts. For example, the input and stored information to be sifted through is of much greater volume than that for chess. Although the number of moves possible in a chess game can be awe-inspiring to a casual player, this 'knowledge space' is much more limited than that of Watson. Moreover, Watson's more general input is relatively 'raw' in the sense that it stays in a 'natural' textual format and it is not pre-coded as the input for chess is. Its goals are also 'richer' in the sense that they are difficult to describe mathematically. Within that three seconds when playing *Jeopardy!*, it has to compose, and then internally evaluate its responses to the cryptic crossword-like clues, rank and weigh various possible candidate answers, and gamble accordingly. These characteristics are what give the problems their muddiness.

The automated playing of another, at first sight perhaps simpler, game than chess, namely poker, reached a new level in 2014–15, when the program called Cepheus was used at the University of Alberta in Canada. Although it presents the player with fewer possible situations to evaluate than, say, even the board game of draughts, poker is high on the difficulty scale because much less complete information is available to players. Full knowledge of game-to-date is not available – for example, the opponent's cards can't be seen, and so – as all who have ever played poker know – uncertainty levels are high. In some ways this means that it's closer than chess to the problems encountered in areas involving, for example, negotiations, political strategies and even some complex diagnostic reasoning in the real world. The famous Princeton scientific polymath and computer pioneer of the mid-twentieth century, John von Neumann[4] referred to this when he wrote: *"Real life is not like [chess]. Real life consists of bluffing, of little tactics...of asking yourself what is the other man going to do?"*

The main approach used by the designers of Cepheus to acquire expertise is not dissimilar to that for AlphaGo or that used by one Arthur Samuel for the game of draughts in 1959 – incidentally in IBM again. With the small amount of computer memory he had available, Samuel implemented a function that tried to measure and maximise the chance of winning for each position encountered during a game. The program recorded the value of the function and how successful the strategy taken was, and it could look a few moves ahead. Samuel got it to play games of draughts against itself thousands of times to enhance its expertise. In Cepheus's case the rules of the *'heads up limit Texas hold 'em'* version of poker were given to it, and it played against itself for two months. Compression and distribution tricks enabled the required '900 core-year computation' to be executed on four thousand processors each playing six billion hands per second for that two months, by which time the system had accumulated vast amounts of data on poker 'moves'[5]. Roughly speaking, this meant that the statistical chances of Cepheus being beaten were so vanishingly small as to be negligible. Now Cepheus is useless when there are three or more players at the table, but in the two-man game, in effect it beats opponents' strategies, although it doesn't actually try to find them and it doesn't even have an inkling of what a 'strategy' is! From an application point of view, what promises to be valuable is its ability to make good recommendations, on things like security policies.

These computing systems and what has been achieved with them are impressive. They are smart and yet they're very limited BAAs. They're not super-intelligent robots who look like Ava in the *Ex Machina* movie, but static boxes of electronic equipment that can run slick algorithms amazingly quickly and maybe access huge information repositories. They are almost solely *playback*. They have been given rules for their operations – that is, programmed – somewhat inflexibly, for their respective special purposes, and have little scope beyond that. For example, even modest extensions of chess, like having a different board, or more kinds of 'fighting men', would need extensive changes to Deep Blue. Some deeper and innovative reflection might be needed. On the face of it, they are also generally seen by us humans as being severely hampered by the fact that they are not embodied, and that they are therefore unable to explore the world using motor and sensory devices to get their own, non-programmed, information, and to have an opportunity to accumulate knowledge and understanding. And this limits their world-views.

However, in thousands of almost unnoticed ways BAAs are changing our lives, and the biggest challenge we have is not to defeat them in games or to make them feel sorry for unfortunate or weak people, or write intriguing novels, but to find the right way to use them. Many interesting questions come to mind when the workings of these systems are studied, in addition to those questions of how they could deal with changes to their given objectives or their rules. Could they be adjusted to deal effectively with inconsistency, error, and deceit? Can you use Watson to answer all the sorts of questions a current affairs panel raises? Or those of poets? Or to (at least) make a stab at answering questions that involve highly complex judgments and relationships or don't have any definitive answers? Does it *want* or *value* anything in particular? How 'good' is its knowledge? Does it reflect on things in general in the way humans do? How 'open' is Watson to new knowledge or to wider dimensions of perception, such as the existence of the supernatural? Did it realise that it had won *Jeopardy!*? Is it *wise* in any sense of that word? Does it have a world-view?

I am assuming that we would have to be able to answer 'Yes' to at least some of these questions if we were discussing an SI. I also assume that it would be intellectually and creatively *ahead* of all human geniuses, such as Albert Einstein, as there are probably fields with which each individual genius wasn't all that well-acquainted. So we're setting the bar high. I'll return to this point later, especially in Chapter 5.

Although the biological and other advances we've thought of might contribute significantly, we'll assume from here on that our platform is a digital computer system. Maybe, in what we might call a *naïve first fix*, as depicted in Fig 2.1, we could put a superstructure over many BAAs – programs like Watson, Cepheus, and Deep Blue, and others – to deal with particular tasks and problems that arise, and incorporate some innovative tricks, or better still, some techniques that only geniuses use, to get to SI? The picture could be much more complex than this, though; for example, a problem that an artificial agent is to perform can be a sub-problem of another more complex problem or task. Solving a detailed tactical problem

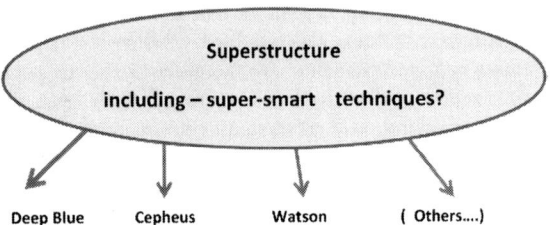

Fig 2.1 A Superintelligence's Superstructure?

during planning, say, for a football match, is a sub-task of an overall game plan, but this is only a part of the task of participating in a knock-out competition, which itself has super-tasks such as plans for development of a team over a few seasons. Given a more general task at a level needing the capability of proposed SIs, perhaps, unlike the development pattern in say Cepheus, Watson or AlphaGo, the task would be given to the machine directly, rather than to a team of human developers, to write the programs and even possibly fabricate the new machine. Coming up with a game plan in very complex circumstances, as in most of life's challenges, the specifications, tasks or problems will be much more 'muddy' than playing any of those games[6]. Moreover, thinking about which problems and tasks to address next would not be easy, and even if we could devise and include adaptive and robust answers to all previously encountered specialist problems, whether developed by humans or machines, 'genius-tricks' needed for novelty are exceedingly hard to identify. Can we put our fingers on what it was that made Einstein and Newton, or Verdi and Bach, stand out from the crowd? We'll look at that question in some detail in later chapters.

World-views for future AI systems? To get some idea of what an SI's values and purpose, goals and openness, objective and scope might be, and the difficulties posed by choosing from contender options, consider again that conversation in Chapter 1 between two, very much in the future, versions of W^*. A clear discontinuity is hinted at in this little fiction – where mankind's 'cleverness' can be surpassed, a haitus that some writers call the (Technological) Singularity. But what evidence do we have that we'll ever get to this point – and are there necessary and sufficient conditions for it to be reached? If, say, W^*25 is the first SI, what genius-tricks did W^*25 have that its 'dad', W^*24, didn't? What might the consequences be,

SUPERINTELLIGENCE AND WORLD-VIEWS

in for example philosophy, science, ethics, maths, politics, and religion?

Let's fast-forward our little conversation a little:

W*21:   But is self-preservation a sufficient goal for us? Is that our final goal?
W*22:   *It's just a step on the way to Utopia!*

Again we hit that problem of Utopia-design. What is the goal of W*22, Sonny, or Ava, and what is the purpose of such developments in general? Is it to 'improve the world', or to 'worship the cosmic mind', or something else? Or maybe it is to ensure human survival, liberty and happiness, although that doesn't seem to register highly with W*22! And the absence of that important question stands out again: What will any SI be able to do that no intelligent human can? For example, will it be able to tell us answers to hard, open mathematical questions such as whether any odd perfect numbers exist? Will it be able to sort out famous paradoxes? What about problems in other domains? In Science there are enigmas about Dark Matter and the dual wave-particle nature of matter, and in Philosophy, a hard question is: Why is there something rather than nothing? In Ethics people have wrestled with the question: Is it ok to kill one to save five? In Theology, there are old chestnuts asking how we square predestination with free will, and how God can be omnipotent and perfectly good in the presence of evil. What about the Big Questions on origins of life, matter and consciousness, or solving more practical problems such as limiting or halting global warming, eradicating world-wide hunger and disease, achieving the perfect community or eliminating all international conflict? It's hard to say what it will do, but if we don't know what we're looking for, deciding whether it's possible and thinking about its engineering is going to be difficult! So that's one of the main issues we need to look at with regard to prospects for the achievement of SI, and their need for world-views.

Suppose the purpose of systems like W*25 and subsequent SIs is going to be to help humans 'prosper or thrive', however we define those terms. Computers will then need to know a lot about us. When we're communicating with each other, we humans take an awful lot of information for granted, and much of it falls into the category of common-sense knowledge[7] that is 'understood' – assumed, and in many cases not normally stated in any formal way. Those future computers must have this knowledge and also detailed knowledge of our work if they are going to help us in that. They need to know what sorts of things we like if they are going to keep us satisfied. To understand our choices and decisions, they must be aware of how we think. To guarantee our freedom they'll have to know what we mean by that word. If they are going to make us happy, or simply look after us, they have to know how we're feeling. In short, they'll have to be aware of our world-views, and possibly they can only do this properly if they have their own, which should then include ours (and our world-views, theirs). So we have the situation illustrated in Fig 1.1, and at the very least we need to transmit to the machines our rich, informal and formal, knowledge of the world, and the rest of our world-views – our beliefs, values, purposes, inclinations and predispositions. They will have to learn from what is known, and knowledge will grow as it is revised, expanded and updated in the light of experience. The computers will need all sorts of knowledge for interacting with each other as $W*21$ and $W*22$ did, and for communicating with humans. The situations they are likely to meet are all going to be at least slightly different and the unexpected can be expected, so while knowledge representation methods should be rich enough to cover known situations and unusual but predictable happenstance, they should not be too fixed. Cross–situational reasoning will need views of multiple 'realms', often in real time.

If we are to take the possibility of SIs ever appearing seriously, they would have to be included in at least some of the narratives we've looked at briefly. With this in mind we're particularly interested here in *how machines acquire their views* of the cosmos and various details of aspects of their existence, like purposefulness and creativity. At the machine level another 'degree of freedom' is added to the three we saw in Chapter 1, viz, the scope for variation between world-view-acquisition-method designs. One option, call it Option 1, is to give such an agent a particular fixed world-view, and the other, Option 2, is to prime it, seed it and let it figure out its own world-view – thereby essentially giving the agent true *freedom*. At the end of the *I-Robot* movie: the robot called Sonny says: *"... Now that I've fulfilled my purpose, I don't know what to do"*. And the response that comes back is: *"I think you'll have to find your way like the rest of us..."*. That's what it means to be free, despite the fact that this would present a problem to the likes of Sonny and maybe people he/it met as he/it found his/its way. Perhaps, like Ava, Sonny's early targets for learning would be limited enough. Ava is not motivated to explore the greatest human achievements, or the seven wonders of the world. She/it wants to start at a traffic intersection.

SUPERINTELLIGENCE AND WORLD-VIEWS

Our second option – prime it and free it – would ultimately mean that agents like Ava and Sonny would *be responsible for* the purpose, values, and perhaps their choices of something very big – some world-view 'Ingredient x' – that they exist for. Clearly the world-views of humans could be affected by any such freedom in artefacts in fundamental ways.

In the 1940s the great computing pioneer Alan Turing, wrote[8] about the possibility of building a 'thinking machine' by putting the 'components' of humans together. The naïve approach – to: *"..take a man as a whole and try to replace the parts of him by machinery"* – seemed to be: *"altogether too slow and impracticable"*. This is somewhat similar to this Option 2 approach, but my suggestion here is that that this approach might not be enough for ultimate success in making 'a human equivalent machine' anyhow, never mind an SI. For example[9]: *"...it is inherent in our idea of a conscious mind that it can reflect upon itself and criticize its own performances, and no extra part is required to do this"*. I'll return to the consideration of the possibility of consciousness in machines later; I just want to repeat here that we can describe the components and detail their functions, inputs and outputs, but this won't produce the purpose and objectives a 'thinking machine' would need. It's clear from the *Jeopardy!* TV show that Watson does have a very fixed and crisp goal. However, it lacks flexibility in how it approaches this and how its achievements are squared with any values. The difficulty scale or 'muddiness' scale referred to earlier, places this above chess, but it would be qualitatively very different from, and in all probability, 'lower' than, the goals of, say, the activities and reflections involved, say, in religion or politics. And an SI would have at least Newton-Leonardo-Bach-Einstein levels of capability. Human geniuses stand out because of their demonstrations of originality, analysis and/or synthesis – sometimes in abstract cognitive space. One of the questions they seem to be outstanding at answering well is: 'What problem will I address next?'. A world-view, including goals and values is needed to address this question. Persistence and a 'frontiersman' or 'intrepid explorer' spirit would also be useful for this of course, and computers can also be very persistent in a sense. But what seems to make most difference in break-throughs is when people appear to have and to use much advanced versions of the sorts of 'shaking things up' tricks we routinely use when trying to innovate. We'll look at some of these in Chapters 4 and 5, and beyond, and consider the difficulty of the question of how genius-tricks might be obtained.

So, in what sort of direction *should* those robots in the movies head? Well, looking at the sorts of things humans choose for 'what to do', there are various options. For example, one that many people would consider to be a suitable goal for a human life is 'seeking *happiness*'. Many biochemists would say that happiness is reached by physical mechanisms, through which appropriate bodily sensations are experienced. But an alternative definition of happiness is to see one's existence in its entirety as *having meaning and being worthwhile* – a 'why' to underpin the 'how', to borrow a thought of the 19th century German philosopher Friedrich Nietzsche. An author trying to finish his/her *magnum opus*, or an athlete trying to win gold at the Olympics, or a monk studying scriptures, might well ignore any negative bodily sensations. At more common and presumably even more transient levels, so might a student writing an essay about something close to their heart or to gain some qualification, or a footballer who is trying to contribute to the village team's performance in the regional league, or a Church member who is studying the Sermon on the Mount for this week's home group. The working, informal definition I want to use for 'true happiness' is: 'being subjectively aware of being comfortable with respect to the big *whys* in one's experience' – presumably in a more persistent and more durable way than in even some of the activities above. Interestingly this fits reasonably with something Daniel Dennett, a Professor of Philosophy at Tufts University, said in his lecture[10] in the TED series. His claim was that to be happy you should get an Ingredient x. *"The secret of happiness is: Find something more important than you are and dedicate your life to it"*.

Returning to our discussion of the new degree of freedom we've just introduced, if we took the option of letting some mechanical agent have its head, we would then have to hope for the best that it would be nice to humans and to humanity. Would this be acceptable to all humans? What if it decided that we'd all be 'happier' if we were doped permanently on 'fun chemicals' such as serotonin? And we can imagine some reactions like: 'What is the relationship between this mechanical agent and the mind behind the universe?' On the other hand, the first of our two options would produce a different outcome. It might allow the world-view of a very smart machine to be given and *controlled by humans*, who we would have to hope are nice people, and the smart agent would be a much more fixed and predictable, although very sophisticated, tool. This could avoid the risk of the 'eliminate humans' threat, but it would, of course, be more limiting in other ways as well, and it would make classifying the artificial system as an 'SI', and maybe even as an autonomous agent, rather suspect. Furthermore, an unfettered SI might be more likely to come up with novel and surprising solutions to many of the problems and Big Questions that humans face even now.

Watson and Deep Blue don't need to have values or a world-view to any degree like the everyday meaning of the term when used of human agents for their finely targeted playing-back. Moreover, such machines don't agonise over Big Questions – those on our cosmic origins and destinations, or on the very large and the very small extremes of magnitude of objects in the cosmos, for example. Self-assessing questions are also irrelevant to them. What are my limits? What do I value or want most? How do I act to achieve my purpose? Are my values consistent with my model of the world, and *vice versa*? They do not ask such questions and don't need the world-views that humans use in order to orientate ourselves in our cosmos or to measure our achievements. Neither of these two systems knew that they had won, or even what is meant by 'winning'! The best modern intelligent machines are only starting to manipulate meanings. To produce artefacts that have the missing features – even only those in our cut-down set of features identified so far – some very difficult challenges would have to be addressed. Understanding and performing everyday operations, never mind doing anything more creative, seems to call for things like *embodiment* and *consciousness*. SIs would probably need the ability to grow an appreciation of the world and experience interaction with an environment, perhaps parallel to the gradual, staged, direct and natural way we humans do, responding to and adapting to changes and to the needs of each set of circumstances. Think of Ava and Sonny again.

This leads to a subject that is of great interest in my own research – how does an agent arrange for, or account for, or live with things that are novel to it in its experience? How can it be *creative*? How does it *explore*? Long ago, Alan Turing wrote[8]: *"in order for the machine to have a chance of finding things out for itself it should be allowed to roam the countryside and the danger to the ordinary citizen would be serious"*. And even then it would still not have had contact with or exposure to things like: *"food, sex, sport and many other things of interest to the human being"*. It's revealing to take a moment to look briefly here at the way computers are said to contribute to more artistic, maybe even creative, endeavours than the functions and scope of Watson and Deep Blue allow.

Now mechanical aids for creative people have been around for a long time. They've been used when 'writer's block' or staleness is experienced. In the world of music, for example, *Musikalisches Würfelspiel*, is a game using dice, said to have been devised by Mozart in the 18th century. A collection of musical measures produced by the human composer, and a set of rules were selected from on the basis of the roll of a die, to generate attractive little minuets. Recently the composer David Cope[11] has produced a computer program, Experimental Musical Intelligence, EMI, to extract equivalents of those 'measures and rules', using computerised techniques, from works by Chopin, Beethoven, Mahler, Bach, and others. EMI breaks the works down into musically meaningful components and patterns, which are used to find frequently-occurring 'phrases' from which to distil a 'style' for each composer, and the system then generates new pieces automatically by mixing and matching these snippets. It seems to be generally well accepted by experts and others that important characteristics of the styles of those composers are, in fact, captured in this way. The output is sometimes quite pleasing or haunting music, claimed to evoke emotions in humans just as, say, Chopin's does. Later Cope scrambled his methods somewhat. Instead of feeding Emily, the successor to EMI, a big pile of historical work of a great master, he gave it a collection of *outputs from EMI* to use as its input. Emily then generated 'its own' musical style. Moreover, Emily could take feedback from Cope, or an audience – and use it to modify the compositions.

Of course some people would raise the valid question: is Emily actually composing? To take an example of an objection, some say that when you are listening to beautiful conventional music, you ask yourself what emotions and thoughts the 'creator' was seeking to capture. I'm not convinced that this is something I personally always do – certainly not consciously – but perhaps I need to have more musical appreciation. Assume for a moment, though, that this sort of communication does go on, but also assume that the machine running the Emily program doesn't have much in the way of experiences of life and feelings. It is sometimes argued that the piece generated then must manifest a certain lack of 'soul' and that its whole *raison d'etre*, 'music as music', and 'art as art', is lost. Perhaps the objectors would want the composition to be less through imitation of outputs, and more through the system being human-like – that is, imitating humans? A 'more sentient' machine might be able to communicate better with people, and move them emotionally as Bach does. More specifically, they'd probably like it to be based on genuine, shared experience.

Another good question arises here: is the new work *original*? Although many will agree that the co-composed pieces output from EMI are recognizable as good music and maybe even 'art', communicating between the computer and the human listener at some level, they do not originate solely within the computer. Emily muddies the water only slightly – it still has to be given substantial user input. This illustration shows how hard it is to even identify what creativity is, an issue I'll raise again, mainly in Chapter 5, and it helps to indicate the difficulty of making creative machines.

Prospects? The prospect of a fairly imminent, say well within 100 years as some predict, superintelligence revolution presents many intriguing possibilities from a utility point of view, and perhaps some threats. But machines with world-views or cosmos-views are still some way off. If they were ever to exist, those world-views would have to be anticipated, and, if so desired, facilitated by humans. We'll have to know or conjecture what sort of relationships the artefacts might have with humans, and the likely impact on our human world-views. However, this will not be easy – *homo sapiens* individuals and societies have been tinkering with their world-views, in particular at the highest (cosmos) level, and often refining them, since the species, with singular ways of thinking and communicating, appeared. We're also a long way from understanding how great human minds step outside the 'normal' scope of human thought or how ingenious reflections take them to works of imagination, creativity and achievement. This is sometimes in abstract cognitive space which the man in the street is often not even aware of. And superintelligence has to improve on that! We're not even nearly at that first base yet!

From a world-view perspective, in addition to the features we've mentioned, our machines will need a list of things that we'll see at the start of Chapter 3 and, as a minimum, emotion, consciousness and maybe even a sense of humour. This brings us to a consideration, at this stage very brief, of the influence, if any, of our third level of agency. The impact of any other-worldly, 'something more important', aspects of human views of the cosmos on machine world-views will need to be considered. In the opposite direction, it will be equally important to consider implications the Singularity would have on our deepest pre-existing views. Looming large here is a higher-resolution kind of recursive dependency like the one we've already noted between SIs and human world-views. The spiritual aspects of human cosmos-views and this particular technology would be tied together, recursively. For example, it is (just about!) conceivable that future SIs could start to see us, their 'creators', as deities, or they could become regarded as 'kinds of deities'. In fact, there are claims that such an entity *might* already exist, and it *might* even have designed our own universe! Let's say, just for argument, that it is possible that such a universe designer exists. Then what? The designer in this case would certainly be, at least, a contributor to the instantiation of our own universe – and in the eyes of many people this would make her/him/it 'a deity'. So the clash with much regular theology and major religions is potentially outstanding. This could result in hostility, which could in time retard technical progress. I'll look more at such issues in due course, but first, we must really get a grip on what world-views are and do. We also need to look at some very brief illustrative examples of views *homo sapiens* specimens have of the cosmos.

References

1. McKenna, P., Rise of the garage genome hackers, *New Scientist* 2689, Jan 2009.https://www.newscientist.com/article/mg20126881-400-rise-of-the-garage-genome-hackers/
2. Gelly, S., Kocsis, L., Schoenauer, M., Sebag, M., Silver, D., Szepesvári, S., Teytaud, O., The grand challenge of computer Go: Monte Carlo tree search and extensions, Communications of the ACM, Volume 55 Issue 3, March 2012 Pages 106–113.
3. Silver, D., Huang, A., Maddison, C. J., et al, Mastering the game of Go with deep neural networks and tree search, Nature 529, 2016.
4. Von Neumann, J., in a taxi conversation with Jacob Bronowski, recalled in *The Ascent of Man*, Episode 13, BBC, 1973
5. Tammeli, O., Burch, N., Johanson, M., Bowling, M., Solving Heads-up Limit Texas Hold 'em, *Proc IJCAI*, 2015. http://poker.srv.ualberta.ca/about
6. Weng, J., Task Muddiness, Intelligence Metrics, and the Necessity of Autonomous Mental Development, *Minds & Machines*, 2009. http://www.cse.msu.edu/~weng/research/MindsMach09.pdf
7. Singh, P, *The Open Mind Common Sense Project*, 2002. Available at http://www.kurzweilai.net/meme/frame.html?main=/articles/art0371.html
8. Turing, A., *Intelligent Machinery*, NPL report, HMSO, 1948.
9. Lucas, J.R., Minds, Machines and Goedel. Philosophy 36. 1961. Reprinted in *The Modeling of Mind*, Kenneth M. Sayre and Frederick J. Crosson, eds., Notre Dame Press, 1963, pp. 269–270; and *Minds and Machines*, Alan Ross Anderson, ed., Prentice-Hall, 1954.
10. Dennett, D., TED talk – *Dangerous memes*, www.TED.com, 2002.
11. Eigenfeldt, A. A., Composer's Search for Creativity Within Computational Style Modeling, Proc Int Symposium on Electronic Art, 2015.

Chapter 3

A Closer Look at Human World-views.

We've been looking at world-views quite a lot already, but I've postponed any deeper consideration of the question of what exactly they are. It's time now to get a better handle on what this term means. Sonny in the *I-Robot* movie I referred to earlier must have had a purpose at the start of the movie, and it had some values and obeyed some well-known 'Laws of Robotics' to help it as it built up its view of the cosmos. It was told that it would have to develop its purpose at least, if it was to become truly autonomous and have meaningful existence. *"I think you'll have to find your way like the rest of us…"*. We'll see in the next few chapters that that wouldn't be easy. For example, dovetailing with humankind in the way that W*21 and W*22 were discussing would be an important consideration in that. For Sonny to fit with us, it would have to know, among other things, what our views were. So to allow us to discuss further what an SI's world-views might be, we need to know something about our own, and I'll be looking at that broadly in the next two chapters.

Views of the World. To sharpen up our working conception of 'world-view' a little we can take the one spelt out by Clément Vidal, Leo Apostel and others at the Free University, Brussels[1,2], prominent contributors to clarifications on the subject who have influenced my own views. Vidal used the following introduction to the term[2]: *"Most people…intuitively have a representation of the world… ,know what is good and what is bad…and have experience on how to act in the world….Every one of us is in need of a worldview, whether it is implicit or explicit"*. Apostel, et al, attempted to make: *"the most profound questions of existence"* explicit. In their publications several questions are posed, representing basic elements that must be accounted for in a world-view, and a simple list or schema is given by Vidal and others[2,3] that we'll refer to, by the abbreviation *A-schema*, quite a lot: *"(a) What is? Ontology (model of being); (b) Where does it all come from? Explanation (model of the past); (c) Where are we going? Prediction (model of the future); (d) What is good and what is evil? Axiology (theory of values); (e) How should we act? Praxeology (theory of actions)"*. Important additional 'second-order' questions are raised about, for example, the sources and nature of knowledge and its scope, so we include *epistemology (theory of knowledge)* in the A-schema.

Answers to these questions are often seen as being what give us our basic coordinates of experience and understanding. We invoke them, consciously or not, for practical things such as exploration, planning and control, and for orientating ourselves conceptually and measuring what we consider to be our achievements. An individual or a community will use world-views when making decisions day by day, for choosing what to do next, and for underpinning '*central* working beliefs' – those that could not be shed without having to abandon many other beliefs as well. For most of us these include the belief that there is a real universe. There are alternatives of, course, like *Solipsism*, which says, roughly, that the 'world' is just a projection of an individual's self and that an individual can't assume that others also have similar cognitive experiences. However, *"...if such a theory were taken seriously it would hardly be possible to take anything else seriously...It is always interesting to find that solipsists..., when they have children, have life insurance"*[4]. I'm keeping away from taking that 'theory' seriously. My working assumption here in the meantime at least is that there is a reality behind the perceived cosmos – some things real and absolute. And our perception mechanisms make it accessible to our built-in cognitive machinery. That reality is the 'top-level-world' – the 'whole scenario' to which we can relate – the whole universe of discourse we're aware of. It includes everything that is philosophically or pragmatically relevant for making sense of life and experience. Happily, belief that the universe makes some sort of sense – that at least some effects have causes, and that there are some generally-accepted norms for reasoning – are also common defaults for central beliefs. Another universal inclusion is the belief that there are other people, quite like us, and, for example, that they also have memories that may often be quite reliable. Getting rid of such beliefs would require a wholesale change in our working pictures of the world.

SUPERINTELLIGENCE AND WORLD-VIEWS

Other beliefs are not shared so widely, for example, specific answers to the Big Questions – such as those on our origins as a species and on the purpose of the cosmos and of life. Individual answers to the question: 'what's life all about?' are many and varied. Lifetime goals and lesser aspirations of individuals and groups affect our practical outlooks, and they are even more varied. Alongside those central beliefs, personality also comes into this picture, as do some other apparently inherent predispositions, traditions, values and 'tricks of the trade', and other methods and patterns of decision-making and even action. Clearly world-views are not simply equated with a choice of lifestyle, but, for someone joining a hippy commune, a suicide bomber, a hermit and an Olympic athlete, basic beliefs, traditions, predispositions, values and other world-view ingredients have important influences on practical matters. Generally world-views can be freely chosen, and it is clear that this is a freedom that should not be trivialised or neglected.

So, individual humans, consciously or not, assemble a coherent, personal collection of values and procedures, ideas and theories, outlooks and objectives, and other bits and pieces, that gives them the basis of a global view of the world that can be fallen back on 'when the chips are down'. It will therefore cover the matters that are of greatest concern to them. It is likely to be largely shared by those in some group, perhaps political or religious, usually being tailored a little for each individual, and its influence can be at a very fine level of detail.

Humans live in complex worlds. I'll be returning to the consideration of what's meant by 'world' from time to time, but, in the meantime, at the level with greatest scope – the top level – which is our primary focus just now, let's take 'the world' of humans to be the broadest physical and mental environment that humans or, for that matter, any human-like agent, can perceive. This perception includes other agents and observers, events, and circumstances.

At the top level, human views of the world involve some sort of representation of, and perhaps an explanation and interpretation of, that world or cosmos, and it also involves taking a position on how to apply that view to life. At lower levels, in some contexts we'll talk about more specialised sub-worlds being 'worlds' as well – for example, 'the world of Sport', or 'the world of Cultural History', or the worlds of Education, or Politics, or Health Provision, or Science, or Philosophical/Religious reflections, or Art. Breaking these down even more finely, there is, for example: 'the world of professional Ice-skating,' or 'the lost world of the Aztecs'.

This means the broadest environment perceived within those reduced settings. Each of these worlds has associated narratives, and the world-views are not all equally good or correct or important. To take an example, people in the same religious community might have very different political, sporting and lifestyle interests, which are considered by the holders to be secondary to, and subject to, their over-arching religious view, and the associated traditional and other values, goals, patterns of action and beliefs. Sometimes conspicuous outliers among world-views receive disproportionate attention, and diversity is particularly evident in such out-lying world-views. Odd-balls are a rich source for this. Some of these people set their lives within a (somewhat narrowly) shared system of 'far-out' beliefs, such as conspiracy theories about things like secret societies that rule our lives, or alien invasions being covered up. Now all of us have minds that tend to look for patterns even if they arise from very weakly suggestive evidence, and we have to be aware that at times they arise from coincidental occurrences. Those who suffer from adherence to conspiracy theories attach undue significance to some chance connections, and try to rationalise them despite the continuing stark paucity or simple absence of any concrete evidence. Their corresponding world-views should be treated with great caution.

We saw earlier that many intriguing questions could be asked about, or even addressed to, an agent like W*25, but it is clear that to answer some of these questions would need specific capabilities that humans have but the current Watson, for example, simply doesn't have. Let's look briefly at some of those very prominent short-comings to start with. *Value judgements* or opinions are needed to answer some questions properly, and for this humans are equipped with a notable functionality – the capability to *evaluate*. Values for humans are preferences among goals or among actions aimed at achieving these goals, and these can be specified at either individual or group level. For example, most humans value 'happiness' which, as scientists, priests and poets affirm, is complex and should not be confused with 'pleasure'. The latter is usually seen as a biological-psychological function of physical sensation, whereas the former can transcend physical limitations or negatives – as when a person rises above a serious disablement, or when missionaries spread their 'good news' to near-inaccessible and inhospitable places, as we saw in Chapter 2. Aspirations and misery, respectively, *do* affect the evaluation of a human's happiness, but the individual's *self* is not *always* placed at the centre of his or her world.

SUPERINTELLIGENCE AND WORLD-VIEWS

That 'degree of freedom' I introduced in Chapter 2 means that under Option 1, W*1, could be *given* some of these preferences by its programmers, but even then would using them be the same as human evaluation? Obtaining and using preferences does not seem to be like this for humans, though. The more exploratory, Option 2, approach to knowledge acquisition would cover much of the value-setting that's needed. Some values are given, but not fixed, being generally received through cultural means, especially, in our early days at least, by transmission from parents to children, and these are contributed to by many complex components and *experiences*. This is a gradual, apparently inefficient and, at first sight slightly weird, process, resulting in an increasingly less provisional frame of reference for helping to decide what is important in the individual's journey through life. Individuals will eventually come to a working knowledge of what they consider as beautiful, positive, or beneficial, for example, and, of course, the polar opposites of these. These attribute values can be ranked, and the *preferences* thus obtained can inform acquaintances about the *sequence* in which that individual will tend to do things. There would have to be quite sophisticated learning and representation features in place if a machine was to get the full benefit of the grounding and provenance of that knowledge of preferences – the justification and experiential history of the knowledge, its sources, confidence levels and derivations. This would be useful, and probably necessary, when addressing some of the tricky questions I listed earlier as humans indulge in deep, reflective and possibly highly innovative musings.

There is a lot more to be said about values, and so much of Chapter 11 is devoted to this aspect of world-views. In the next few sections we're going to look a little more closely at what that term 'world-view' itself means for humans, and the nature of some of the other features of world-views. As we saw in Chapter 1, some of the evaluations that humans do could *contribute to* their world-views, and in identifying and assessing values we arguably also need to *make use of* world-views. This is an example of the kind of *recursive interdependence* that we've already noted.

As well as its lack of the ability to evaluate, in a certain sense, any current machine also lacks other capabilities, and one of these is *Purposefulness*. This might be considered alongside values and actions in the A-schema, as being very importantly linked to our human views of the cosmos, and it might be wondered how much purposiveness, if any, Watson has. It would be hard to detect it or otherwise prove its existence by a detailed examination of its individual components. The much-advanced computing facility W*25, might display to humans say, a personalised view of its world and its place in it, and to possess considerable originality of sorts, and perhaps an ability and a freedom to make judgements. As we observed it and tried to comprehend it, we might note that it displays a certain purposefulness in its operation (eg self-preservation), and this might consolidate our impression that it has a high form of 'awareness' and agency. Indeed we would then see a discernible gross conception of the agent, taken as a whole, but we might not be able to actually find it in its components, or deduce it from them alone.

This all links in with issues of *openness* and *identity*. The extent of openness to two arguably critical influences – science and technology on the one hand and the wider spiritual elements we met briefly above and in Chapter 1 on the other – appears to be central these days in separating out the sorts of world-views people have. Major relationships exist between our own personal networks of structures and any widely shared narratives and world-views we encounter. And we should be open to and use narratives that are well grounded in objective and subjective facts and any wider master narratives.

The question of the agent's identity also looms large. Oxford Professor of Pharmacology and neuroscientist, Susan Greenfield[5], has claimed that we can see identity as 'subjective mind states'. Context, including values and memories, is incorporated into a still wider framework: *"a narrative of your cohesive past-present-future"* – your personal narrative. Common sense, anecdotal evidence and the relatively small amount of more systematic evidence that's emerged recently all clearly show that we lean on memory as we build future scenarios. Greenfield made the point that a developing identity depends on context – and it does so more and more as time goes on. So if someone had grown up with wolves in the wild, his or her identity would not be as it is in their current circumstances. In particular the society to which they belong has a great influence on identity, and how they're seen in it is a key feature. The way that society and the world sees them also develops over time, and it can be seen in narrative terms, as a *sort of* journal of their consciousness. So it is very difficult to give an account of a life without using narratives.

Together with those 'bigger' narratives of science and religion, social, political and economic narratives have big influence on our world-views. For example how the economy that we are born into develops is hugely influential, as can be seen easily by comparing a child in the developed world with one in the third world. And it's

SUPERINTELLIGENCE AND WORLD-VIEWS

important for us to think about our place in the changing natural world. The wide range of responses to the environment, including many associated with those other narratives, and with emotions such as wonder and awe, can be triggered by an environmental narrative. Furthermore a technical narrative in its own right rather than as an add-on to a scientific one, and one of our main interests in this book, is a big, very dynamic, influence on our histories. It consists of an unfolding, ever-developing, collection of technical accomplishments. This has been called the: *"the (modified) milieu in which (man) lives"* and which *"is no longer his"*[6]. Artificial technical necessities pull us around in ways that are in a rapid state of flux. Incidentally, Greenfield's book is about some aspects of the ways in which technology can change minds, brains and outlooks, and we'll look at some of these briefly in Chapter 10. To help us in our endeavours to make sense of the world as I've mentioned, and for solving hard problems, narratives do make use of logical arguments, but they also require, among other things, emotions such as empathy and values and a willingness and an ability to live with riddles, uncertainty and even ignorance – there are so many mysteries that we will have to leave some unsolved. A sense of humour may help in facing up to the imponderables of life. Fundamentally responses to narratives are needed for innovative, broad and creative thinking, as we'll see in Chapters 4, 5 and 13, for example.

We looked briefly at some suggested narratives for the future, and I pointed out earlier David Deutsch[7] for one says that we can't ever get an SI. He extolls the unique capabilities of humans, although he does not attribute their uniqueness as due to anything non-natural, such as the well-known concept of *Imago Dei*, that humans mirror God in their three-fold spiritual composition of intelligence, emotion and will. He says: *"There can be no such thing as a superhuman mind...No concepts or arguments that humans are inherently incapable of understanding"*. When he talks like this about humans he presumably means mankind collectively. Incidentally, I haven't seen anywhere where he presents his explanation of the evolutionary leap that's said to have happened less than 70,000 years ago[8] from the smartest non-human animal to the human, or the chasm between the smartest genius and the man-in-the-street. My guess is that he does not rule out the possibility of the equivalent of a genius-trick, or similar, having produced a small advance to get to human level capability, analogous to any leap there would be from pedestrian human level intelligence upwards. Yet, interestingly, somehow he believes that he can rule out the possibility of something parallel or similar happening again, at those higher levels.

Historically, non-natural explanations and wider narratives which incorporate, for example, the concept of *Imago Dei* mentioned above have had influences on all this sort of speculation. They are, of course, very interesting, but they need careful handling by the layman. Frank Tipler who wrote a lot about the science around the Omega Point Theory, was very influenced by two theologians, Teilhard de Chardin and Wolfhart Pannenberg whose speculations, while stimulating, raise deep questions. For example, could we human: *"universal explainers..."* as Deutsch classifies us, rely for our persistence on mankind's own provision, and could a move towards SIs lead to worship of something strange?

Cosmos-views. A human individual's world-view, at the top, 'whole world', level, is the beholder's 'big picture', and it can provide an apparatus to assist in getting a deep working understanding of reality – hopefully, a reasonably dependable means of prediction and control for basic purposes of life and of more abstract contemplation of principles that may govern it. The purposefulness mentioned above, with its frequently somewhat fuzzy goals, is often linked to the fact that the people concerned may see the ultimate whole as being much greater than they are. We're reminded of what Daniel Dennett said in that lecture in the TED series I referred to in Chapter 1, that to be happy you should get *something more important than you are* to live for. Ideally this topmost world-view, including an Ingredient x that you can commit yourself to, should cover or qualify *everything* of importance to us – our complete universe of discourse. Does Deep Blue have 'something more important' like this? Perhaps the status of chess as a game has to be upheld, or humans have to be shown to be inferior to machines? I don't think so.

From now on I'm going to use the term *cosmos-view* to distinguish and capture the scope of this topmost-level view of the collection of things in their entirety. G. K. Chesterton has written about the fundamental importance of such a wide-angle view, which could have practical benefits[9] *"...the most practical and important thing about a man is still his view of the universe....the question is not whether the theory of the cosmos affects matters, but whether, in the long run, anything else affects them"*. In addition to this cosmos-view, and possibly within it, there is, as noted earlier, typically a series of 'lesser, component world-views' for various sub-worlds, such as those for sport, lifestyle, health, or religious activities. They should all fit harmoniously with the 'global schema', the cosmos-view. We'll look at key Environmental, Technological, Scientific, Philosophical and Religious components in Chapters 9 and 10.

In Sigmund Freud's[10] very, very ambitious, very, very high level, definition of world-view that we met in Chapter 1 he referred to: *"an intellectual construction which...leaves no question unanswered"*. At any given time on the journey towards this 'stretch target', we would like the 'answers' obtained using the view to be close to our final answers to our questions, but even if this is not so, they can often serve as 'working answers' in the meantime. I would like to broaden the concept out a little by including *emotions* and *will* as well as the intellect in its scope. The complexity of the human intellect in offering mental abilities and processes related to knowledge is paired with a complex system of emotions. *Feelings* add to the 'control levers' available to us for navigating through our lives as we'll see in Chapter 13. Human reasoning ability varies from person to person, and it is accompanied by a similarly varying capacity to control and use emotions – especially in the sphere of inter-human relationships. *Will* also comes into the picture in relation to purposefulness, to account for an individual's freedom of action, driving us towards particular actions. Prominent manifestations appear, for example, in intentions to make a difference to one's circumstances, in accordance with, but possibly limited by, the pursuit of some goals or causes.

So cosmos-views result from an apparently common tendency to establish an all-embracing, but not always consistent or well-understood, interpretation of things that are experienced and thought about. And remember that patterns of action, patterns in phenomena and values are included in assessment of the world – how reality is perceived. In some ways, having a cosmos-view is like putting on tinted goggles. It affects how all things in the cosmos are perceived, grasped and understood, no matter how they are materialized or otherwise have substance in reality. Within it and because of it, at lower hierarchical levels, the other specialised, domain-specific world-views are often given a wider perspective but also a particular slant. In the opposite direction, it seems reasonable to assume that if one of those key component world-views is absent, the value and possibly the validity of the global world-view – the cosmos-view, the complexus – could be compromised to some extent. If the view is overly constrained in any way, it could be like losing a dimension in the spatial world. If one uses only a 2-D mapping of a 3-D world, the overall perception is likely to be deficient for some purposes. For example, a neat solution to the problem of drawing three lines to produce a set of four equilateral triangles from a given initial equilateral triangle is easier to find if the third dimension is used.

The influential post-war American philosopher, Nathan Goodman[11] says that our theories and models are the only reality that is accessible to us. So, he says, we make worlds by creating, refining, and manipulating symbols. Despite the fact that this philosophical idea of world-making is somewhat alien to the man-in-the-street, we can learn from Goodman's conceptualisation of 'worlds' (or our world-views and cosmos-views) as heuristic, abstract models constructed by scientists, the man in the street and others, to help us, if we can harmonise them, and perhaps distil from the collection a decent cosmos-view, to get a handle on life. *"The physicist takes his world as the real one, attributing the deletions, additions, irregularities, emphases of the versions to imperfections of perception, to the urgencies of practice, or to poetic licence"*. On the other hand: *"For the man-in-the-street, most versions from a science, art, and perception depart in some ways from the familiar serviceable world he has jerry-built from fragments of scientific or artistic tradition and from his own struggles for survival"* Goodman goes on to say that the man in the street world*"...is the one most often taken as real; for reality in a world, like realism in a picture, is largely a matter of habit"*.

How do we choose between cosmos-views? Goodman discusses a number of standards and criteria for good, measured, realistic world-making in his sense of this term, and says that they can be wrong. *"...while readiness to recognise alternative worlds may be liberating, and suggestive of new avenues of exploration, a willingness to welcome all worlds builds none.....A broad mind is no substitute for hard work"*.

Vidal[2] gives three criteria for choosing between contending world-views. Summarising drastically, the first is the *objective* criterion, for looking at things like the view's internal logical consistency and simplicity, and the scope of the view – roughly speaking its coverage in terms of the number and variety of cases that it pertains to. The *subjective* criterion includes how well the view fits with personal experience and knowledge, how well it promotes a rewarding approach to life, and how much it invokes emotions. The third criterion is an *inter-subjective* one, which is concerned with minimising conflicts between agents, promoting the interests of a group as a whole, and the use of narratives in communication. To help with the use of the criteria, various assessment tests were proposed. *Fact-value tests* deal with how *what is* co-exists with what *ought to be* the case. For example, political oughts, such as 'we ought to cut back on manufacturing industry's carbon footprint', triggered by global warming, can be somewhat inconsistent with other existing oughts, such as that there should be jobs for all. *Value-action tests* are about consistency between, say, values and concrete decision-making and action. In *Fact-action tests* the only value of interest is efficiency. Other examples of tests are: the *we-I*, and *it-I* tests for world-view fit with the interests of the wider community, and compatibility with the best modern knowledge, respectively. The *we-it* test asks whether society as a whole is compatible with what we know about the objective world. The A-schema is probably the 'state-of-the-art' framework for making such choices at present. For example, we could use the case study presented by Vidal to get a 'fruitful Science-and-Religion dialog'. This could potentially lead to either a religious cosmos-view being adjusted as a result of scientific discoveries, even in cases where matter, time and space considerations are of secondary importance, or to a scientific cosmos-view with 'wider' perspectives added in, extending its scope to include, for example, values and perhaps a theory of action.

I roughed out a three levels of agency picture at the start of Chapter 1. Humans, at the lowest of the three levels, can be seen as sitting above the level of the 'smart' machines and BAAs that are available now, like Cepheus or Emily, but very definitely below any SI successors, as we'll see in later chapters. If Dennett's 'something more important' were to be an agent, it could conceivably be at a higher-than-human-level of agency, and, arguably, also above the SI level. For example, our top level, supernatural agent in that agency picture would fit the bill. I will consider what we can know about the nature and attributes of such a 'higher level of agent' from time to time in later chapters. As I'll also point out throughout this book, the world-views of human agents and the possible development of mechanical agents which possibly have world-views are linked recursively in very interesting ways. Both have/would have 'something more important' to be dedicated to. Would this 'something' have to be the same in both cases?

So we're back to what we were considering in Chapter 2 – the most interesting question about all this in our present context – where would any super-smart machines fit in to our world-views? These are sometimes referred to as artificial intellects or *artilects* – Hugo de Garis's term for an artificial intelligence superior to any human's in *at least one* domain of knowledge which has the will to use the intelligence. Strictly speaking, our 'SIs' are generalised artilects. Our question is: could machines like these really lie between those other two levels of agency? We'll look at this question later, but first of all, let's repeat our working definition of superintelligence from Bostrum[12]: *"By a superintelligence we mean an intellect that is much smarter than the best human brains in practically every field, including scientific creativity, general wisdom and social skills"*. Taking such an SI to be an agent implemented on a digital computer, and ignoring the word 'practically' for our purposes, we take it that such an SI would be intellectually *ahead of* all human geniuses, such as Albert Einstein. And the SI would need human-quality-plus world-views for this...

References

1. Apostel, L., and Van der Veken, J., *Wereldbeelden. Van fragmentering naar integratie*. DNB/Pelckmans, 1991
2. Vidal, C. Metaphilosophical Criteria for Worldview Comparison, *Metaphilosophy* 43 (3), 2012 http://homepages.vub.ac.be/~clvidal/writings/Vidal-Metaphilosophical-Criteria.pdf
3. Aerts, D., Apostel, L., Bart De Moor, Hellemans, S., Maex, E., Van Belle, H. and Van der Veken, J., *World Views. From fragmentation to integration*, VUB Press, 1994. See http://www.vub.ac.be/CLEA/pub/books/worldviews.pdf
4. Bell, J. S., *Speakable and Unspeakable in Quantum Mechanics*, Cambridge University Press, 1987.
5. Greenfield, S., Mind Change, Penguin Random House, UK, 2014.
6. Ellul, J., The Technological Order, in *Technology and Culture*, Proceedings of the Encyclopaedia Britannica Conference on the Technological Order Vol. 3, No. 4, John Hopkins University 1962.
7. Deutsch, D., *The Beginning of Infinity*, Penguin Books, 2011.
8. Harari, Y. N., *Sapiens*, Harvill Secker, 2014.
9. Chesterton, G. K., *Collected Works, volume 1*, Ignatius Press, 1986. See also, An Introduction to "Heretics" by Chesterton, G. K.http://www.biblebell.org/visitors/heretics1.html
10. Freud, S., In New introductory lectures on psychoanalysis Lecture XXXV: 'The question of a 'Weltanschauung'. In *Abstracts of the Standard Edition of the Psychological Works of Sigmund Freud*, Carrie Lee Rothgeb, editor,1933.Retrieved. fromhttps://www.marxists.org/reference/subject/philosophy/works/at/freud.htm
11. Goodman, N., *Ways Of Worldmaking*, Hackett Publishing Companv, Inc, 1978.

12. Bostrum, N., *How Long Before Superintelligence?* http://www.nick-bostrom.com/superintelligence.html[Originally published in Int. Jour. of Future Studies, 1998, vol. 2] [Reprinted in Linguistic and Philosophical Investigations, 2006, Vol. 5, No. 1, pp. 11-30.]

Chapter 4

Illustrations of World-views.

Sometimes it's easier to furnish an example of a concept, a thing or a class of things, rather than to describe it fully and correctly, or to define it in a principled way. So I think I should give a few examples or versions of real human cosmos–views, in various levels of detail and emphasis, to complement what's already been sketched out about what they are like. The concept is of something that is very fundamental and which works its way, often without being acknowledged explicitly, into all aspects of the lives of the holders – when they are making decisions and taking action, for example. Threads that come from science and theological narratives, social, technological, economic, environmental and political narratives, and personal narratives interweave with, for example, narratives that focus on sport, family, education, art, history and hobbies. I will not attempt to assemble a comprehensive collection covering every one of the huge varieties of cosmos-views that can be found. The aim of the informal assemblage I'll present is simply to illustrate cosmos-views of different *flavours* in our sense of being open or closed, considering the material only or not, and perhaps being 'evangelical' or otherwise intentionally transformative or not. It should be clear how those flavours are manifested within the examples, and hopefully there will also be some glimpses of *kinds* of world-views – for example, environmental, political, religious and technological world-views – and some idea of the values, beliefs, visions and activities of the holders of the views we've chosen to look at.

Pleasant Outlooks. Cosmos-views are often quite complex even in simple-looking cases. The virtues of the first approach to life I'm going to describe were energetically proclaimed by one William Cobbett, whose life straddled the turn of the 18th and 19th centuries.

SUPERINTELLIGENCE AND WORLD-VIEWS

He was very active politically, advocating a 'root and branch' adjustment of contemporaneous outlooks and patterns of life as a means to re-establish the relatively 'Arcadian' England of his youth – one that would not include, for example, the ubiquitous paper-mills and other factories that sprouted up later in his life when the industrial revolution had hit country-dwellers hard. In his view, people's lives could be seen as a sort of collaboration with nature and the land. His message thrived at the time, and even today self-sufficient small-holders refer to the 'gospel' he preached. The spectacular full title of his most famous book[1] summarises that message: *"Cottage economy; containing information relative to the brewing of beer, making of bread, keeping of cows, pigs, bees, ewes, goats, poultry, and rabbits, and relative to other matters deemed useful in the conducting of the affairs of a labourer's family; to which are added, instructions relative to the selecting, the cutting and the bleaching of the plants of english grass and grain, for the purpose of making hats and bonnets; and also instructions for erecting and using ice-houses, after the virginian manner"*.

This title-paragraph gives us some insights into the life-style of cottagers around Cobbett's time. For their world-views, most values and a substantial chunk of their kitbag of aptitude, knowledge and know-how were handed down from parents, grand-parents and other relatives, and subsequently expanded and tailored by experience through learning and experimenting. Cobbett sketched out some idea of the life-style and attitude associated with their shared cosmos-view. On a day-to-day basis this meant hard work: *"...I have known many men dig thirty rods of garden ground in a day; I have, before I was fourteen, digged [sic] twenty rods in a day, for more than ten days successively"*. A man at Portsea was reputed to have dug: *"...forty rods in one single day, between daylight and dark"*. A rod is about five paces, or five and a half yards, so we're talking about a lot of cultivation. In fact: *"the well-fed healthy man"* working as hard as this would have quite a bit over for 'farm-gate' sales or barter. Income could be supplemented from the fruits of animal husbandry, which went together well with arable work.

Cottagers' knowledge went beyond propositions, being grounded firmly in the 'real', man-in-the-field, causal experiences, events and entities that lay behind them. They developed a genuine respect – perhaps even a love – for the countryside where they lived, and they would have had a sense of being part of a landscape continuously in flux as climate and cottager continually influenced it. Some poets have waxed romantic about this oneness with nature. In his books of poems[2, 3], the Irish Nobel laureate, Seamus Heaney, wrote about a perception of a 'good life' experience in something as simple as crossing a field. Like a cottager in the past, we presume, Heaney had a deep, conscious and deliberate sense of engaging with and enjoying the flora and fauna. Cottagers or small farmers would be tuned in to observing an: *"empty briar...swishing"* or simply the: *"...overhang of a birch, grass and seedling on the quarry face"*, and that: *"Heather and kesh and turf stacks reappear, summer by summer, grasshoppers and all"*. The appearance and behaviour of birds and animals would be intriguing to them, like: *"the perfect eye of the blackbird (that) watched"* and tracking: *"...a hare until the prints stop just like that. End of the line, where did she go"*. The poet captured the sense of oneness cottagers would be likely to have had with the land when he wrote about the physical country and the mental country being 'married'.

Hunter-gatherers in ancient or present-day primitive societies had or have perhaps even more constrained looking cosmos-views to those of the cottager or small farmer. Details of their environment were or are of the utmost importance to them, warranting very meticulous scrutiny and providing deep 'understanding' and knowledge of importance to them. As with our farmer, we should be very careful not to have too low an estimate of their knowledge, skill, industriousness and cleverness in meeting their daily needs. With little of the ready access to technical and aesthetic literature widely available today, they still used or use quite sophisticated reasoning and evaluation mechanisms. In some ancient societies[4], for example, any piece of flint an ancient tribesman found was said to have a special 'personality' – hard, flaky, and: *"..willing to behave in a predictable, ultimately useful way when struck by a piece of horn"*. The same 'personality' was manifest in all lumps of flint found, and this led to the allocation of its class label – viz, 'flint'. Interestingly explanation of the characteristics of things like flint frequently came in the form of a myth, often having a mystical character. However rules such as: *'if flint then flaky'* were used hard-headedly and successfully for practical purposes in *modus ponens*, or other valid argument forms and rules of inference, just as we use 'laws of nature'. So there was some propositional and rule-like knowledge. Of course, it could be supplemented in various ways, and, some of these may have been simply ways of overcoming gaps in their knowledge that have in many cases been bridged in more advanced, worlds. Those 'ways' would, presumably, make possible for these people, like a small farmer or cottager, a sort of contentment, based on having a non-volatile, wide-angled view of their world and their existence, in harmony with their means of subsistence. Ancient or primitive people sought or seek coherence and meaning in their views of life and the cosmos, just as much, and possibly more consciously than we do.

On the other hand, wide swathes of the world were or are inaccessible to such communities, and therefore might as well not exist *for all practical purposes* (FAPP). There is therefore, despite all of the 'gap-filling' extravagance of their myths, for example, a conspicuous net parsimony in the world-views that support survival in even modern hunter-gathering societies, as is captured succinctly by the philosopher Daniel Dennett[5]. He is prompted to marvel about a conversation he had with one of his students: "*...about his efforts on behalf of a tribe living deep in the Brazilian forest*". Dennett wanted to know whether they knew about the main participants in the cold war at that time, and the student replied that:"*There would have been no point in it. They had never heard of either America or the Soviet Union. In fact, they had never even heard of Brazil!*". Dennett's marvelling came from the fact that: "*It was still possible in the 1960s for a human being to live in a nation, and be subject to its laws, without the slightest knowledge of that fact*".

Such tribesmen almost certainly have fairly extensive social and religious conventions to fill out their, assumedly common, relatively fixed and long-lived, conceptual frameworks, but they would have no need for some other *kinds* of world-view such as the wider political aspects of views that we probably maintain to some degree. Daniel Dennett[5] has a view on why this is astonishing: "*...it is because we human beings, unlike all other species on the planet, are knowers. We are the only ones who have figured out what we are, and where we are, in this great universe. And we're even beginning to figure out how we got here*". We'll be looking in due course at how confident Dennett was entitled to be in his assessment of such collective human knowledge. However, the fact that knowers and knowledge creators can be satisfied, and can sometimes live unusually fulfilled lives with relatively cut-down world-views, can sometimes even be seen in modern Western society. For example, even amongst those who are comfortable with, and who have benefitted from, the 'figuring out' Dennett refers to, there are those who have deliberately *dropped out* to follow *the good life* to some extent.

Less Pleasant Circumstances. Now what about the views of people in more miserable settings? What about the cosmos-view of someone in a death cell, possibly tortured beyond the point where even minimal balanced reflection is impossible? Or a heavily overwrought 'sweatshop' worker in some primitive society, or a hunger striker, or a missionary in some pressing but inhospitable and dangerous circumstances? What would the world-views of a Brazilian street child be like, or those of someone whose world had been devastated by a war that was now over, but which had left chaos, deprivation and exploitation in its wake? What would the cosmos-view of a refuge-camp inhabitant after some strife or another look like? Such 'less pleasant' outlooks must be widespread. Consider the many hundreds of thousands of displaced citizens housed in holding camps in various countries as I write. These are not holiday camps – for example, hunger is common and hygiene is poor. Dreamy poetry appreciating these states of affairs is, understandably, light on the ground. It is salutary to think of what the world-views might be of the many other displaced people in hot spots around the world. How do they compare to those of, say, a Heaney or a Cobbett? Many will be educated, civilised folk who through no fault of their own have been caught up in some uprising or invasion. Have they retained all of their values, and are their formerly useful hand-me-down and other know-how of any use now? Survival is again a key contender for being their main driver FAPP, at least temporarily – and it possibly swamps aspects of their previously-held cosmos-views.

David Livingstone, the 19th century Scottish explorer and missionary to Africa exemplifies yet another sort of outlook, one which often led to deliberate self-denial and discomfort in ways that most people would, I think, say is impressive. His life gives a good example of how a person's cosmos-view dominates their thinking and deliberate acting, and can produce a drive to transform the world. It also provides a story that illustrates an open mind – pragmatically in his frontiersmanship and exploration, and theologically in his great emphasis on his 'something more important'. Livingstone was by any standards a phenomenon. He was gifted in many ways, and he matched this gifting with an amazing work-rate and many important personal character qualities, some of which helped him keep moving ahead in his exploration, mission, healing and other work in very difficult circumstances, often while beleaguered and in dire straits. Other gifts allowed him to work in harmony with the native African population, despite the fact that he was very resolute in his ideas, having a strong Scottish personality. Like the cottagers he did not merely give assent to the truth of basic propositions, but apprehended what was in his view truth, in a way that captivated his

SUPERINTELLIGENCE AND WORLD-VIEWS

life and penetrated his world-views at all levels. He led some of the most significant and most dangerous scientific and exploration expeditions of Africa that were seen in the nineteenth century, sponsored by the Royal Geographical Society. One of his motivations was social – to lay bare the scourge of the slave trade, that persisted even then, some years post-Wilberforce, in order that it could be cut it off at source. But most important to Livingstone, from the writings of William Blaikie[6] which are: *"chiefly from his unpublished journals and correspondence in the possession of his family"*, remained the possibility of bringing the 'good news' of 'Christianity and civilisation' to as many as he could. His foremost driving objective was his religious missionary zeal to bring a message to people who had not been exposed to it.

I do not want to rehearse any more details of Livingstone's work, which are well covered by Blaikie, and how he continued to work tirelessly in his continuously-adaptable direction of travel, with worsening health, driven by his all-embracing cosmos-view. As a summary, I'll simply use some words of the American reporter, Henry M. Stanley, a correspondent on the *New York Herald*, who famously met up with Livingstone in November 1871 in Ujiji near Lake Tanganyika, saying "Dr Livingstone, I presume?" He was greatly impressed by Livingstone in the four months and four days he lived there with him: *"To the stern dictates of duty, alone, has he sacrificed his home and ease, the pleasures, refinements, and luxuries of civilized life. His is the Spartan heroism, the inflexibility of the Roman, the enduring resolution of the Anglo-Saxon – never to relinquish his work, though his heart yearns for home; never to surrender his obligations until he can write FINIS to his work"*.

The features of the cosmos-views of Livingstone that stand out are without doubt related to what Dennett called 'something more important'. Driven individuals can of course also be found in non-religious directions, and they can have 'conversions' that change their lives in a way similar to religious conversions. This is not the place to go into great detail, but there are abundant good illustrations of this. The mention of one such famous driven person will suffice at this point. The author and intellectual Arthur Koestler, a great literary figure in the mid-20th century, had a personality and an education which led him to come to believe very deeply in the intellectual order he found in Communist teachings. His commitment to and zeal for his convictions led the extreme experience of living on death row, waiting in solitary confinement to be shot. He was released before that could happen, but, interestingly, not before he had had a sort of epiphany – what he saw as a mystical experience of some sort in his cell, leading to a complete turn-around in his convictions and his cosmos-view. Of most interest here is the fact that he made the discovery that: *"a higher order existed, and that it alone invested existence with meaning"*[7]. He referred to ultimate reality as being transmitted by: *"a text written in invisible ink"* that: *"the founders of religions, prophets, saints and seers had at moments been able to read"*. He was a trained scientist and that remained a key interest, and he also wrote on a variety of topics such as capital punishment and anticommunism, but his new beliefs after conversion seemed to be swayed towards psychic phenomena, in contrast to Livingstone's Christian faith. His ideas on the future of the cosmos seem to have been a bit like Tipler's – all about merging at death into the cosmic consciousness. He was Vice-president of Britain's Voluntary Euthanasia Society, and when his body was found beside his wife's in 1983, the notes that were left led to the belief that the deaths were by suicide.

Genius World-views? The biographies of many other high achievers also reveal similar drivenness towards some clear purpose as for Livingstone, or some changing purpose as for Koestler. What about the cosmos-views of those who would be by the widest common consent called 'geniuses'? These are clearly interesting here, given our working definition of SI. Well, it is true that drivenness is often a characteristic of geniuses of various kinds. In common parlance, by 'genius' we could mean, among many other things, a scientific superstar, or a great entrepreneur or an inventor. This word has even been used when referring to a great sportsperson or 'celebrity' with a special entertainment prowess. The illustrative geniuses I consider here are those who are far in advance of the crowd intellectually and creatively. They stand out because of their demonstrations of originality and insight and cognate skills in scientific or artistic pursuits. Exceeding the achievements of the rest of humanity must be partly a function of physical inherited characteristics and gifts bestowed free of charge on geniuses, with genetic material arranged in precisely the right patterns to impart such things as high intelligence, insatiable curiosity and enthusiasm[8]. These gifts are outside the control of the individual. But what is clear is that just, say, being in the top decile in IQ is unlikely to be enough. Vision, dedication, resoluteness, motivation and opportunity for development are needed. Many people conceive a genius as being inspired or possessed by a divine power. The view of others is that genius comes from natural endowments. Either way the genius is equipped to go beyond the normal to achieve outstanding results.

Some acknowledged 'geniuses' have given an inkling of their wider world-views, and for example, their knowledge acquisition activities frequently interact with personal values and aspirations and cultural factors. Those world-views are not what make them geniuses, but they certainly contribute to the selection and representation of domains and problems and to the evaluation of the results of creativity. So appropriate world-views are necessary, but not sufficient, for genius. While Livingstone and Koestler are not geniuses in the very narrow, but admittedly rather imprecise sense I'm using the term, many geniuses have equivalents of the respective ideologies those two expressed, especially those concerning the way society and the world in general should be run. Their values and purposes are every bit as conspicuous as for Koestler and Livingstone, and there are indications of strong innate psychological drivers such as curiosity and setting the bar high for personal satisfaction. They often exhibit a desire for consistency and coherence between ideas and activities, and I am not ruling out wider and more practical concerns such as some pursuit of prestige, power and material prosperity. They sometimes express views on the nature of reality and what we can know about it, and the influence of domain authorities and presentation norms are often acknowledged, although these certainly do not bind true genius. All these factors reticulate and interact with each other, and operate within domain-specific, national, spiritual and, for practical purposes, even local communities. The resulting world-views of the geniuses come into the creative processes at two stages according to Emeritus Professor of Mathematics at the University of Oxford, Sir Roger Penrose [9]. One of these he calls 'putting up' – *"...forming judgements, so that only those ideas with a reasonable chance of success will survive"*–and at the later 'shooting down' stage, where those ideas that were put up are evaluated and judged. So although they are not sufficient, I strongly believe that history shows that the world-views are necessary.

Before looking at two prominent illustrations of geniuses and their world-views, I want to look very briefly at Galileo Galilei as an example of a man of very complex world-views. He was a genius in physics and astronomy. He had creative ideas in both the arts and the sciences, although his contributions in the latter were arguably very much the more significant, and certainly better known. However, he engaged in debates about sculpture and painting, and he was an expert member the Academy of Drawing in Italy, in *chiaroscuro*, or perspective drawing. This was considered in Galileo's time to be a science, being concerned with the accurate use of illumination and shadow. It could also be used in painting and drawing to create a sense of 3-D, especially with complex geometric figures. Galileo also produced a series of watercolours depicting the mountains and craters on the Moon. This spelt trouble for him with the theological authorities at the time, who held that the moon was a perfect sphere of radiant vapour. Galileo's infamous controversy with the church led to him being put under house arrest in his old age. But he was, in fact, an affirmed Christian believer who, like Francis Bacon whose life also straddled the 16th and 17th centuries, saw a large overlap and strong interrelationship between religion and science. Bacon was happy to have the familiar dual view of God being the author of the book of Scripture and the book of nature. Similarly, in a Letter to the Grand Duchess Christina of Tuscany in 1615[10] Galileo suggested that this was perhaps what Tertullian, the early African Christian apologist, meant when he wrote that God is known through both nature and revelation.

Let's now focus on some illustrations of what we know about the world-views of Einstein and Picasso who would commonly be regarded as being geniuses. I don't wish to imply that these world-views are 'more worthy' or 'more true' in some sense than those of the man in the street or any of those above, as there is a danger of over-revering the world-views of outstanding characters. For example in the case of artistic genius, theologian Francis Schaeffer[11] writes: "*...the greater the artistic expression, the more important it is to consciously bring it and its worldview under...judgment...The common reaction among many, however, is just the opposite*". Here I simply want to pick out and acknowledge any impact their world-views appear to have had their outputs, with a view to judging whether or not this is relevant to SI-genius. The rule of 'best endeavours' must apply here as it is very hard to find direct evidence of geniuses expressing what their world-views were in detail, and that's why I restrict my ambitions to drawing illustrations from 'genius domains' where the genius qualities are perhaps somewhat better documented than others. The chosen domains are science and art.

I first consider the domain of science, moving on some centuries after Galileo's contributions were made. Although scientific creativity seems at first sight to be restrictive due to its focus on largely hypothetico-deductive methodologies, and formal and disciplined constraints on the endeavours that lead to originality, this will certainly give us some idea of at the very least a possible lower boundary of what's needed if we are ever to specify more general artificial geniuses. Discovering new explanations or improving existing explanations by putting them together requires imagination of the highest order. It is inherently creative[12]. Think of Galileo's stepping away from the 'given' explanation of heavenly bodies as relatively small celestial balls of incandescent vapour. Concepts of huge solid spheres traversing huge distances and orderly, systematic routes are not something Galileo derived purely 'from the outside' through experience – by encountering them in his day-to-day life in Pisa. Albert Einstein picked up on this *plus ultra* approach to discovery when he outlined[13] his take on the nature of genius – and not only in science – in his tribute to Emmy Noether, an exceptional mathematician who died quite young. He refers to her relatively brief personal narrative, her attitude to life and the use she made of her talents. Einstein writes: *"....there is, fortunately, a minority composed of those who recognize early in their lives that the most beautiful and satisfying experiences open to humankind are not derived from the outside, but are bound up with the development of the individual's own feeling, thinking and acting"* [my emphasis]. He lays out some thoughts on the world-views of geniuses in general, describing people who through some special gift, are concerned with pursuing the: *"most substantial and desirable end to be achieved"*. He also lets us know some of his own values and views, and objectives, as he refers to people like Noether: *"However inconspicuously the life of these individuals runs its course, none the less the fruits of their endeavors are the most valuable contributions which one generation can make to its successors"*.

Einstein was certainly aware of his own strong contributions, but he was frequently self-effacing[14]: *"the contrast between the popular estimate of my powers and achievements and the reality is simply grotesque"*. He also shows some of the things he valued most in life when he expressed a 'passionate' sense of social justice, but contrasted it with his more reserved, private detachment from social ties. Politically he was a democrat and stood for equality. He put a high value on the pursuit of understanding for its own sake: *"...it is not the fruits of scientific research that elevate a man and enrich his nature, but the urge to understand, the intellectual work, creative or receptive"* and as he emphasised this in his eulogy for Noether. Einstein explained his motivation as: *"...Feeling and desire"* which are, he said: *"the motive forces behind all human endeavour and human creation, in however exalted a guise the latter may present itself to us..."* However, he referred to having a sense of there being a 'something more important' in his widest-angle world-view: *"The individual feels the nothingness of human desires and aims and the sublimity and marvelous order which reveal themselves both in nature and in the world of thought."* Einstein talked of the desire to have a holistic experience of the universe, and in his view: *"...it is the most important function of art and science to awaken this feeling and keep it alive in those who are capable of it"*. This was related to his personal work ethic based on devotion and hard work, and his view of what enquiring minds are all about – *"...a deep conviction of the rationality of the universe and what a yearning to understand, were it but a feeble reflection of the mind revealed in this world"*. He explicitly extended this to other domains where genius can be manifested. For example, he admires those like Heaney who can increase our appreciation of the world. He says that they: *"hold the mirror up to it by the impersonal agency of art"*. The work of Heaney's fellow-countryman George Bernard Shaw had, he said, a: *"liberating effect on us such as hardly any other of our contemporaries has done and have relieved life of something of its earth-bound heaviness"*.

On war, culture and the state he had very definite views that could be seen as functions of his Jewish background and the times he lived in. He was very proud of his background and he acknowledged its influence on his thoughts. *"The pursuit of knowledge for its own sake, an almost fanatical love of justice, and the desire for personal independence—these are the features of the Jewish tradition which make me thank my stars that I belong to it"*. He, largely as a result, put a high value on the sanctity and pre-eminence of life, and all things sacred. His background impacted on his creativity in the scientific world: *"the Jewish tradition also contains something else, — namely, a sort of intoxicated joy and amazement at the beauty and grandeur of this world, of which, man can just form a faint notion. It is the feeling from which true scientific research draws its spiritual sustenance..."*. So this derivation *was* to some extent 'from the outside' complementing inner experiences like those of Noether and Galileo. Those inner experiences of geniuses could result from inputs of so-called 'luck' and some inspiration or possession by a divine power as referred to by Plato[9], but it is unlikely that we would find anything useful in that direction that would be of help for SI designers. Perhaps we could random mutations or fluctuations to represent the 'luck', but they could not be guaranteed to be advantageous or even positive.

Now, although Einstein acknowledged genius in diverse domains, there are differences between them. For example, in art a new output: *"rarely proves an old one wrong"*, as often happens in science[12]. The German Philosopher Wilhelm Dilthey whose life spanned the 19th and 20th centuries distinguished between the natural sciences and art, and he highlighted the influence of a genius's world-views on his or her creativity. He identified different "classes of life-manifestations", one of which he called "expressions of lived experience", through for example language, and he referred to 'great' painting, for example, going beyond the creator's self-revelation[15], as its: *"spiritual content is liberated from its creator"*. What is of interest here is that there is a relationship between any expression of genius and the reality of: *"the manifold forms in which a commonality existing among individuals has objectified itself in the world of the senses"*. Those 'forms' cover high pursuits such as philosophy, art, religion, politics and science, but also the work-a-day activities of life, FAPP. So such expressions of genius are closely related to an individual creator's world-views, and also to the world-views of communities to which he or she belongs. For example, art allows an: *"inner experience,"* which helps: *"bring me to a consciousness of my own individuality. I experience the latter only through a comparison of myself with others"*– linking with 'the outside'.

As in science, imagination-driven exploration and discovery in some artistic domains such as poetry, but perhaps especially music, is undertaken within very tight rules of engagement, but it still requires creativity[13]. Pablo Picasso, is widely acknowledged as being one of the greatest painters of the 20th century, but his particular form of genius stretched to other artistic 'expressions of lived experience' such as sculpture, ceramics, and even stage design. We learn a lot about his world-views from a book[16] by Gilberte Brassaï, Picasso's photographer, in the preface to which the American writer Henry Miller captures how Picasso formed his appreciation of the world which he tried to present to others: *"We discover that Picasso is an omnivorous reader, au courant to everything taking place in the world of letters as well as the world of art, to say nothing of the everyday world which, through Picasso's eyes, seems crazier than ever"*. Picasso had clear goals, and a statement of his about the necessity of planning is often quoted by motivation speakers. Moreover, like a scientist, he seemed to be seeking something that was already there, a sort of ideal, a search for perfection: *"...from one canvas to the next, always go further and further"*.

Picasso once said: *"I always aim for likeness. A painter has to observe nature, but must never confuse it with painting. It can be translated into painting only with signs. But you do not invent a sign. You must aim hard at likeness to get to the sign"*. The 'translation' here is the part of the process that's 'not derived from the outside'. However, he mocks the *artiste peintre* – the professional visual copier – whose attitudes and drives contrast with his own, and elsewhere he questioned what reality was. Both art and science are, almost by definition, partial expansions. *"...No metaphysical formulation can have more than relative validity because it attempts to arrive at a totalization that transcends experience"*[15]. However exploratory probes, along with sense data from experiences and activities, shake up our world-views, leading to:[15]*"asking, believing, presuming, claiming, taking pleasure in, approving, liking and its opposite, wishing, desiring, and willing"*. Together art and science may lead to personal 'rules of engagement' being established and purposes and values being posited. And when we humans have a purpose, we often start to seek and expect fulfilment.

It has been said that each of Picasso's works was a response to something he'd seen or felt, often something that surprised and moved him. There were many cultural, natural and social influences on his work. Old masters released his creative energy. He said that he liked Mathias Grunewald's Crucifixion, and wanted to 'interpret' it his way, but what he ended up with was something new. Picasso also had a love for and appreciation of aspects of nature in which he found inspiration for humble materials like stones from the beach. Further inspiration came from unconsidered trifles of civilisation, such as the contents of wastebaskets, and Brassaï says he was like Leonardo in this. From Brassaï we also get some insights into Picasso's motivation: *"He wants to confer to his every movement a historical value within his history of man-as-creator. He wants personally to place each of his acts within the great annals of his phenomenal life, before other people do it"*.

Received horizons, narratives and world-views can all be rebelled against. Einstein 'ganged his own gait', and Picasso believed that the world's *status quo* need not be his. His social and political values, impacted by his Spanish origins, influenced him in his work: *"We must fight fascism wherever it manifests itself"*. With regard to other values, he appreciated the prestige and success that arose from his work: *"It's often been said that an artist ought to work for himself, for the 'love of art' that he ought to have contempt for success. Untrue! An artist needs success. And not only to live off it, but especially to produce his body of work. ..."* However, his higher values put money *per se* in the shade: *"What's the point of having even*

SUPERINTELLIGENCE AND WORLD-VIEWS

more money when you already have enough? You can't eat four lunches or four dinners just because you're richer".

To summarise some characteristics and apparent world-views of genius that are manifest in some of the above example comments of Einstein and Picasso, although it is possibly a little more oriented to Einstein than Picasso, we can paraphrase Pinker[17]: geniuses immerse themselves in their visions for many years, absorb huge amounts of questions and answers, develop many strategies for tackling difficulties, and keep an eye on what others are doing. They're mindful of their place in history and the esteem that their work brings. They choose the right problems, maybe by 'lucky' chance. However it's clear from the comments of Einstein and Picasso above that they also generate their creations internally, supplemented by 'external' sketches, etc, of course, invoking their world-views in the generation of their conjectures and using them in the evaluation of their ideas against them, at what Roger Penrose calls the 'shooting-down' stage.

In childhood and education, then, the care-givers and other influencers of potential geniuses have a real challenge as they prepare for and widen their scope. What would stimulate the genius who created Newton's Laws, and many other things, or, Shakespeare who created much of the English language, or Mozart or Da Vinci? Ellen Winner of Harvard University has sought for[8] the psychological and personality characteristics of individuals who have seminally and permanently changed what goes on in their particular domains, and highlights early starting, individuality, and an obsessive interest in something along with an exceptional aptitude for it. To revolutionize they must be born at the right time, and historical and cultural factors come into play, plus a lot of circumstance and providence. Creators are willing to sacrifice relationships, comfort and time. Their talent and creativity of course stand out and they tend to be very different from the herd. Their goals are ambitious, and their purposefulness exemplary. This is accompanied by self-belief, a welcoming and thriving on competition. They are open and non-conforming – they: *"gang their own gait"*, and they are not markedly risk-averse.

The world-views, if they could exist, of a cockroach or of an amoeba, would be unlikely to have any deeper 'something more important' features that we've seen in most of our illustrations of world-views. And it is clear that it would be very difficult to give world-views to a machine in order to get to W*25 and beyond. But they are necessary, if not sufficient, for genius and therefore for superintelligence. To contemplate installing them in our machines, and consider what they might look like, we need to begin from some understanding of human cosmos-views. These brief sample illustrations may help a little to determine what the specification of this particular potential product of agency – genius – would have to be like. They also illustrate the difficulty of even getting such a specification. From these non-systematically selected exemplars we can see why automated world-view generation and management, especially those needed for geniuses or SIs, are clearly still some way off.

References

1. Cobbett, W., *Cottage Economy*; John Doyle, 12, Liberty-St New York, 1833.
2. Heaney, S., *Field Work*, Faber and Faber, 1976.
3. Heaney, S., *Seeing Things*, Faber, 1991.
4. Jacobsen, T., Mesopotamia, in *Before Philosophy*, Frankfort, H. and H. A., Wilson, J A., and Jacobsen, T., Penguin Books, 1949 (originally published U of Chicago Press, 1946).
5. Dennett, D., *Freedom Evolves*, Penguin Books, 2003.
6. Blaikie, W. G., *The personal life of David Livingstone*, John Murray London, 1880.
7. Koestler, A., *The Invisible writing: Autobiography 193153*, London, Collins/Hamish Hamilton, 1954.
8. Winner, E., *Gifted Children: Myths And Realities*, Basic Books, 1996.
9. Penrose, R., *The Emperor's New Mind*, Oxford University Press, 1990.
10. This text is part of the Internet Modern History Sourcebook. http://history.hanover.edu/courses/excerpts/111gal2.html
11. Schaeffer, F. A., *Art and the Bible: two Essays*, Hodder and Stoughton, 1973.
12. Deutsch, D., The Beginning of Infinity, Penguin Books, 2011.
13. Einstein, A., Obituary, New York Times, May 5, 1935. Retrieved from: http://www-groups.dcs.st-and.ac.uk/history/Obits2/Noether_Emmy_Einstein.html
14. Einstein, A., The World as I See It, originally published in Forum and Century, vol. 84, pp. 193-194, in the Forum series, Living Philosophies. Also published by Simon Schuster, 1931.
15. Makkreel, R., "Wilhelm Dilthey", *The Stanford Encyclopedia of Philosophy* (Summer 2012 Edition), Edward, N., Zalta (ed.), http://plato.stanford.edu/archives/sum2012/entries/dilthey/ Referencing various other works.

16. Brassaï, G., *Conversations with Picasso*, Translated by Jane Marie Todd with a Preface by Henry Miller and an Introduction by Pierre Daix, University of Chicago Press, 1999.
17. Pinker S, *How Mind Works*, Penguin Books, 1997.

Chapter 5

Thought Experiments on the Computer?

Computing systems can certainly contain digitised memories and they have potential to generate dynamic internal 'visions' of the future. These capabilities would be needed if they were to 'remember' episodes in the past and possibly 'chew over' events in order to try to 'make some sense' of them. Different strategies and algorithms could be tried to help to explain earlier happenings and to predict the results of different actions.

Humans do this very well – when thinking about past sports events, political transformations and personal experiences, for example. We can also think ahead – try out different strategies mentally in various scenarios to see what is most likely to bring good outcomes. The cleverest people can conduct complex and novel thought experiments, and it is interesting to ask if it might be possible to arrange for machines to do likewise, or at least help human thinkers in their thought experiments. Thinking of how machines could approach this capability shows both the richness of humans in this respect, and the difficulty of doing this within computers.

As a backdrop to our thinking on this, consider an exploratory practical study my colleagues and I devised a few years ago on the use and scope of various computing techniques such as learning and reasoning techniques. To summarise, the main objective of the experiment was to acquire and use the rules that governed the observed behaviour of an adaptive autonomous agent, a robot investigator, which had some prescribed goals that the observer did not know. For this we used a mobile robot as the observed agent[1]. The robot's mechanically generated actions and behaviour were observed by a second computer system, representing a researcher. The observer 'reflected upon' the observations automatically, mainly using machine learning techniques, in order to assess their relevance to some known objectives, and subsequently to avail of any such potential practical value the ensuing insights might give the observing system.

In the course of the project we showed that, by using motion information that is acquired from video analysis or a visual system, which we used for convenience in place of direct observation of the robot, the observing system is able to reverse-engineer the recorded behaviour of the robot investigator agent successfully. It can thus uncover to some degree the underlying algorithms and the rules of behaviour of the observed robot, known to the (human) experimenters, but not to the observer system. The observer system was able to utilize its approximations to these in order to predict a specific successful action for given scenarios.

We designed a test harness for use in such experimentation and called it the Experimental System Apparatus (ESA), which consisted of a number of key modules. In the study an important feature was that the investigating robot's knowledge, K', so discovered, could be compared directly with the 'unseen' generating knowledge, K, in order to evaluate the computer techniques used. Another module was used for prediction, taking advantage of the K' patterns learned from the action analysis module to anticipate pertinent decisions of the observed robot agent, for use by the observing system to meet *its* goals. The success of this prediction could be readily assessed and it gave an additional validation of the newly discovered K'.

Why am I describing this project? Well, the ESA is seen as modelling, for example, what goes on in some two-agent episodes that occur in 'the real world'. It could emulate a *second order predator* (the observing system) by analysing image sequences of movements of a 'predator' (the observed robot) and the 'prey', whose behaviour patterns were being observed. Both of those observed entities have activities governed by a set of known, predefined behavioural templates. The results of the investigation gave 'proof of concept' verification that the emulated second-order predator was able to identify action patterns of the 'prey' and the 'predator' correctly, and that it could predict the 'predator's' decision using its approximations, the K', to the pre-defined algorithms and rules, K. This learning scenario has much intrinsic interest, but the researchers on the team believed that the techniques and many of the insights obtained were transferable to other domains, especially those involving time series mining, such as Business Processes and Surveillance.

For our present purposes just imagine, as our own thought experiment (*about* an 'automated thought experiment'!), that all of this capability was exercised 'under the covers' – internal to the 'observing system'. Using ESA, various trials – for example to choose which videos to 'recall', and to look at alternative algorithms for both learning and generation of 'futures to try' – could be carried out *purely internally*. Which videos of the opposition football team should be 'recalled' coming up to a big match? These might show that a 'pre-conception' or 'assumption' was wrong in the light of the data – eg someone who has a reputation as a great defender is not so good defending on his right side. Perhaps scenarios could be assembled to 'imagine' what would happen in certain circumstances. Unsupervised automated generation of animations from text has been possible for some time[2] – for example semantic keywords are extracted from a story in standard ways and used to search an image database for images that can be ranked, and stored or output. Could combinations of such techniques be used to get a view of future consequences, and so resemble thought experiments? Choices of algorithms and actions could then be made as appropriate to meet the purposes of the observing system.

Creative Thought Experiments. In the present context I use this example of experimentation to hint at how it might be possible to approach what can be seen as a *sort of* digitised human thought experimentation on machines. The basis of 'mechanised thought experimentation' could be available through the use of videoed episodes and other data, in combination with the 'cognitive activity' of the emulated observer. It gives a hint as to how deep reflective thinking on, for example, ensembles of moving entities can be carried out by computers, as sorts of thought experiments, and eventually by autonomous robots, or even by SIs.

What sort of thought experiment am I talking about? Einstein has frequently been quoted as saying in a speech in Kyoto on December, 1922 that he was sitting in the patent office in Bern when a sudden thought experiment made him think of a man falling but not being able to feel his weight. This led to the theory of gravity. After some years of deep thought this intuition led to his conclusion that gravity is not in fact a force of attraction, growing with the distance between objects as Sir Isaac Newton had intimated in his mechanics. Now remember that Newton's was the only coherent explanation of gravity available to the young man in that patent office in 1907, or to anyone else at that time. It was a good theory in that it accounted for the observed 'classical' behaviour of falling objects on Earth and also nearly all aspects of the interaction of heavenly bodies. But Einstein discovered that no such gravitational force of attraction is required. Newton's apple fell to Earth because of curvature in the fabric of space-time. Gravity is maximal where space-time is most curved, and the mass of the Earth made a good dent in space-time so his apple was falling into that valley. Some people summarise Einstein's theory of general relativity, of which this is the main insight, by JA Wheeler's phrase: *"matter tells space-time how to curve, and curved space-time tells matter how to move"*.

This wasn't mere thinking about an experiment, or simply imagining experiments that Einstein wanted to carry out in reality and not just in his imagination. This exercise was dreamed up because it is impossible to run the desired experimentation in the real world, and it was aimed at illustrating and elucidating very complex and intangible phenomena. Einstein was able to go on to his theories of relativity from his thought experiments. Many decades later, he referred to that when he wrote about[3] another thought experiment based on him chasing moonbeams or sunbeams conceptually, that led to a fundamental conflict with experience and existing theory. Now I do not want to go into the debate that this triggered[4], but use this illustration to see what we can say about of the self-expressed mental creative processes of the genius, and envisage the effects of one particular genius' activities and pronouncements on his or her first 'audiences'. I hope the analogy that I see with the robot experiments is clear, and that it is not too stretched.

Genius in Different Domains. Now we saw in Chapter 4 that genius-level contributions from different domains are not directly comparable. For example, for a work of art it has been suggested by Francis Schaeffer that: *"...there are four basic standards: (1) technical excellence, (2) validity, (3) intellectual content, the world view which comes through and (4) the integration of content and vehicle"*[6]. There no consideration of *falsifiability or refutability* as seen in scientific theory – the inherent possibility that it can be proven false. According to Schaeffer[5], fine art is not analysed or: *"...valued for its intellectual content. It is something to be enjoyed"*. But if you take a look at say a collection of Picasso's art – say via the images under his name in Google – it's hard to see that word 'enjoy' as having its standard meaning. Maybe productions such as mother and child or landscape paintings are 'easy on the eye' but some would suggest that a scientific discovery such as the first law of thermodynamics could be 'enjoyed' every bit as well as *Guernica* by the uninitiated! Other aspects of evaluation are also difficult to specify.

Does the *value* of an output in any domain reside in encouraging the evaluator to see the world in a new way, or is it in extending knowledge horizons or enabling enhanced quality of life and even more measurable things like wealth creation? As an aside, for this reason I personally use a rough principle that I believe is self-evidently needed when assessing how creative contributions from different domains or sub-domains are: the innovativeness of contribution or creativity of any given output should, to the extent possible, be measured, assessed or compared in the light of any prominent bigger sub-domain to which it belongs, and it should be evaluated in the light of any broader 'rules of engagement' in which they are expressed. Using Bostrum's definition of SI means that their would-be designers are really forced to look at genius across domains, despite the inherent difficulty of matching this entails.

Getting artistic and scientific genius in one agent will be very hard! Consider fine art. The disposition a creation invokes in the inner experiences of the beholder originates from signs from 'outside', and any help they can give is useful. The experiences of another individual – eg the painter – will often be very different to the observer's. The immediate appeal or otherwise of the 'beauty' captured in one of Picasso's works might catch my attention deeply, but I get a fuller understanding of the work from his recorded comments, and reported actions as well as that final output. He also left clues on the way he arranged his exhibitions, and he developed his paintings in sketches as well as the finished works of art such as The Demoiselles. The creator's other outputs and the creator's worldviews, including known predispositions, values, goals, beliefs and other outlooks should ideally also be considered. Of course scientists don't create *ex nihilo* and even painters use what's already there, as we'll see just below, but in both cases we'll see that creation is involved in the sense of making something that did not exist before. In the case of science the word 'make' may not be wholly appropriate, and even for fine art it's not totally true. But in both cases – and in the latter case in particular – a bit of thought shows that in, say, a painting or a mathematical equation – *The Demoiselles*, or $e=mc^2$ – something was produced that did not exist before.

I may use different sorts of analysis in the understanding of the outputs of scientific and artistic geniuses, but all creative work has to obey certain 'rules of engagement'. Usually, that is, because one characteristic of true genius is the propensity to break out of the normal patterns of thought and activity. However having established 'rules' enhances appreciation by the initiated, and it allows the next generation to take advantage of bequests. These rules vary from *rigid* in Science, through *tight* in music, and *careful* in literature, to something *more heuristic*, sometimes closer to 'free-form', in visual art. We all operate under a common humanity – general human nature, but source agent and receiver agent have their own individuality. The receiver has to suppress some idiosyncrasies and put emphasis on 'alien' experiences and their expression to project into the source's world-views if inputs from the genius are to be experienced. There may also be helpful aids around which experts and commentators have designed to help us in our understanding, and constraints on the validity of our appreciation of the work due to the 'language' used in the dialogue between source and receiver personalities. This can only be done up to a point. It must be remembered, though[5] *"...that just because an artist – even a great artist – portrays a worldview in writing or on canvas, it does not mean that we should automatically accept that worldview. Art may heighten the impact of the worldview, ...but it does not make something true."*

We appreciate geniuses, but we are unlikely to understand them. To emulate their capabilities, or surpass them on machines, would mean that we would have to understand them better than they understand themselves, as there is much that is unconscious in acts of creation, as we'll see. Immanuel Kant has a lot to say about genius. He writes that[6]: *"an author...himself does not know how he came by the idea ...nor is it in his power to devise such products at his pleasure"*. He does not know how to pass: *"his procedure"* on. And eminent, genius level, creativity that's considered 'great' in its field is qualitatively different from the more everyday 'creativity' exhibited by professionals or other experts solving problems in their own discipline and by the man in the street. Again thinking of visual art in particular, we saw in Chapter 3 that Nathan Goodman says[7] there seems to be a *sort of* default 'real world' from which the genius draws input, for: *"...reality in a world, like realism in a picture, is largely a matter of habit"*. Attempting straightforward copying, perhaps for information purposes, is reminiscent of Picasso's views on 'visual copiers' seeking to capture reality through straightfoward likeness that we saw in Chapter 4.

However, what the outputs of genius are and, to the extent possible, how they are generated, are of rather more interest here than what is invoked in the beholder. Thought experiments might be of help in generating such outputs. Part of the reason I have introduced thought experiments is that they give a stage upon which genius can be exhibited, in other domains as well as science. If we could harness these for use by human geniuses, and perhaps digitise them, we would take a small step towards realising some desirable SI capability. Kant tells us[6] that *artistic* genius is a natural endowment, which 'gives the rule'. Here I take 'the rule' to be the 'rules of engagement' mentioned above. Unfortunately for those aspiring to be SI designers, such rules say nothing about *how* the ultimate artistic outputs are produced, but govern aspects of *what* to produce, identifying the models that are worthy of emulation by other artists. Interestingly, the artistic genius is thus a rule-maker: *"Genius is the talent (natural endowment) that gives the rule to art."* and is not only, according to Kant, simply a rule-breaker – something that some other people would say is necessary for genius. Kant says that while genius is a natural endowment, it is in the grip of powers beyond the control of the genius, who cannot: *"devise such products at his pleasure"*. The powers included in artistic genius are: *"imagination and understanding"*. And to this must be added *taste* which *"like the power of judgment in general, consists in disciplining (or training) genius"*. Kant assumes that the judgement of fine art invokes mental faculties that are like those involved in judging beauty.

Artistic creative talent is different from the sort possessed by great scientists. Among other things, for visual artists the presentation itself is an important part of the creative activity. Brassaï, Picasso's photographer[7] says: *"the very act of 'presentation' is an important moment in his creative process. It is through other people's eyes that his work becomes separate from him, that his mind becomes aware of what he wanted, what he succeeded in doing"*. Perhaps we could copy Emily's methods to produce outputs somewhat 'like Picasso's', but the problem of seeing an output through an observer's eye, which is analogous to listening to Emily's output, remains. Furthermore, Kant says that the activities of the scientist can be described: *"...start from distinctly known rules that determine the procedure we must use in it."*, and that they don't need to use the way: *"aesthetic ideas ... are offered or expressed"* that artists need. I don't think Einstein would agree with the first part of this statement. Scientific genius requires great originality. While *"... science progresses while art does not"* [8] a simple derivation procedure is not enough for this progress – for example when something that the scientist has never experienced is conceived of and transformed into knowledge. Galileo managed to do this. Moreover, Picasso held the view that 'newness' was really an illusion in the case

SUPERINTELLIGENCE AND WORLD-VIEWS

of fine art: *"Whatever you might say or think, you're always imitating something, even if you don't know it"*. Anyhow, while there is a big difference between purely imagination-driven exploration and partially experiment-based scientific discovery, it can be argued that in both science and fine art creativity, both are necessary. David Deutsch[9] says that theories in science are: *"...guesses – bold conjectures. Human minds create them by rearranging, combining altering and adding to existing ideas with the intention of improving them"*. Contrary to the widely held idea that science is predominately about experimentation, the main use of such experience, says Deutsch, is actually to: *"choose between theories that have already been guessed"*. Both highly disciplined and formalised creative activities and 'more open' creativity use the approaches of trying new things and transforming old things, as we'll see.

Is there a Mechanics of Creativity? So how much can we get on any actual mechanics of genius creation? It turns out that the answer is 'not enough to write an SI program'! In a letter to the French mathematician Jacques Hadamard, Einstein tried to set out, as best he could, how he 'created'. Importantly he said that:[10] *"The words or the language, as they are written or spoken, do not seem to play any role in my mechanism of thought. The psychical entities which seem to serve as elements in thought are certain signs and more or less clear images which can be 'voluntarily' reproduced and combined"*. The words and logic are still needed, later, however, to allow communication to others. He says that those psychical entities are, in his case, *"...of visual and some of muscular type. Conventional words or other signs have to be sought for laboriously only in a secondary stage, when the mentioned associative play is sufficiently established and can be reproduced at will"*, and that play has to be *"...aimed to be analogous to certain logical connections one is searching for"*.

The connection with the investigating robot experiments described at the start of this chapter is stretched but it's hopefully clear, at least for scientific genius. Could there be an equivalent of the 'digital experimentation' described that uses a robot to support 'experimentation' like Einstein's 'visual and muscular' imaginings or Picasso's sketches? The great scientist's 'canvas' was internal. Einstein's conviction was that his desire for 'discovery of a universal formal principle' and his 'ten years of reflection' eventually produced a driving initiation for that full thought experiment. But what made him ask the specific questions he asked? There is obviously a major first step needed to get the seed question to initiate any experiment, and this is beyond our robot system. We'll return to this question, for example in Chapter 13. Einstein used thought experiments in abstract space for trial and error – his 'blank page' equivalent of Picasso's canvases or other media.

In order to get a better idea of how the imaginations of an SI genius might proceed, let's now look further at artistic creativity in painting as against that which, say, scientists, and even musicians manifest. Deutsch[8] has noted one important difference. *"Mathematics has its proofs and science has its experimental tests; but if you choose to believe that [some artist/composer/poet/...] was...inept...then neither logic nor experiment nor anything else can contradict you,"* [my insertion]. Looking at brush strokes won't explain great paintings. Neither will a study of its composition or organisation, colour, texture, or shape. Indeed, apparently neither art nor science at the genius level can be reduced to data, evaluated mechanically, quantified and ranked. Something of that sort of processing was used for Emily the 'mechanical composer', but it is very unlikely that data can produce an 'Aha!', or even the sorts of seeds of the ideas that Einstein came up with. That's no more likely than the marvellous new 'interpretations of objects' in Picasso's eyes being obtained by data analytics. Brassaï once wanted to know whether Picasso's ideas come to him: *"by chance or by design"* and Picasso responds[10]: *"I don't have a clue. Ideas are simply starting points. I can rarely set them down as they come to my mind. As soon as I start to work, others well up in my pen. To know what you're going to draw, you have to begin drawing...When I find myself facing a blank page, that's always going through my head. What I capture in spite of myself interests me more than my own ideas"*. Elsewhere, as we've seen, referring to his reworking of Grunewald's Crucifixion he said that he had tried to 'interpret' it but had ended up with a completely new painting. Brassaï quotes an acquaintance as saying on another occasion: *"Picasso reacts to what's around him. Each of his works is a response to something he's seen or felt, something that surprised and moved him"*. Picasso gives us some inkling about his own creative take on everyday things when he comments on the fact that Michelangelo saw images in marble. He says: *"I understand how you could see something in the root of a tree, a crack in the wall, in an eroded stone or pebble. But marble? It comes off in blocks and doesn't evoke any image"*. I'll return to this just below, but here I want to highlight the problems for data analytics in this.

Interestingly, Picasso worries away 'muscularly', to use Einstein's term, at something until he gets it right. He does this physically, on paper for example, rather than by the pure thinking that Einstein did: *"I too often tell myself: 'It's not quite there yet. You can do better.' I can rarely keep myself from redoing a thing – umpteen times the same thing. Sometimes it gets to be a real obsession. After all, why work otherwise, if not to better express the same thing? You must always seek perfection"*. This 'perfection' is a sort of ideal – what Deutsch calls 'objective beauty'. Picasso seems to mean an objectivity that seeks to identify some universal truth or meaningful patterns or analogies within what is observed and express it in an expert way. He gives an insight to this via Brassaï: *"How do you like this tiny little slip of a woman?...I've redone her I don't know how many times"*. He was apparently looking for something that was objectively 'there', meeting some level of beauty or standard of implementation that could only be attained after going through many 'hard yards'. Of course, creative scientists and mathematicians often have similar experiences, as we'll see in the case of Cédric Villani in Chapter 13. There could be *mistakes* that need attention, and there could be standards to be aimed at in innovations, since some are *better* in some sense than others.

What could we mean by 'better' in fine art? Perhaps closer to perfection? But as we don't know *ab initio* what perfection is, how do we compare versions? It is not as simple as saying something like: people like complete, continuous images rather than broken up ones, or they cherish symmetry, or pronounced contrasts or 'happy' colours, much as, say, rats appear to like harmonious music better than mere noises. If there is a petal clearly missing in a flower it often irritates, but progressives might say that it stimulates the beholder. There might be some cultures and some people, with distinctive cosmos-view, including beliefs, values, and innate predispositions, that would appreciate or be moved by that omission. But if it's innovative like Picasso's cubism, for example, such 'parochial tastes', as they are referred to by Deutsch, would hardly give satisfactory explanation – via 'the rule'? – of the 'improvements'. Perhaps a theory of 'betterness' like the simple one Nelson Goodman[7] used, or more simply still, some crude formulation of what is beautiful to humans, like the happy colours referred to above, is data-dependent in a relatively trivial way. The theory is limited by the fact that the range of human senses must limit the scope of appreciation that the artist seeks to communicate.

The unpredictably of Picasso's art is clear. But, is it due to 'randomness effects', or is the unpredictability due to the deep impenetrability of reality? In science and mathematics, there is built-in unknowability. Quantum mechanics, chaos theory and even things like the inconsistency or incompleteness theorems in logic limit what can be asserted. So could artistic taste be seen as a property of matter, fitting with the laws of physics and chemistry, or biology as in sexual attraction in animals, or in the attractiveness of blue objects like straws and bottle-tops to a male Satin Bowerbird as he tries to catch the eye of a mate? Or should it be left as unjudgeable or subjective? According to people like David Deutsch[8], attractiveness might depend on some sort of 'objective beauty' to get replicated through breeding. Deutsch hopes that: *"when we better understand what elegance really is, perhaps we will find new and better ways to seek truth using elegance or beauty"*. He talks about us having new senses, new qualia, beauty of new kinds and new experiences – mañana! We have some way to go to get the 'how tos' for creativity in painting and other modalities.

I want to point out now the importance of world-views in fine art. For illustration, consider Brassaï's description of a broomstick that Picasso has burned some marks into, being: *"...struck by his infallible gift for giving life to any material he touches. From the first stroke, he guesses, invents, and reinvents the most fitting technique, as if the sources, secrets, manual skills, age-old experiences of all the graphic and plastic arts have always been instantly at his disposal"*. Brassaï says Picasso liked the work of the Japanese artist Katsushika Hokusai because he shared much with him: *"...a keen curiosity about every aspect of form; the power to capture life on the fly and fix it with a fluid, concise stroke; patient attention; dazzling execution"*. Hokusai also experimented. *"He sometimes used extra-pictorial means, the tools at hand — for example, the tip of an egg dipped in ink. He liked to improvise, make humorous, comical, tender, or cruel pastiches"*. Another incident shows how Picasso saw wonderful variety even in the mundane, in: *"..a box full of stones, bones, fragments of plates and crockery that have been ground by the sea, all engraved and sometimes carved slightly"*. They were: *"modelled and polished by the sea, engraved and etched by his hand"*. Picasso says to him: *"The stones are so beautiful you want to carve all of them. And the sea shapes them so nicely, gives them such pure, such complete, forms, that we have only to add a finishing touch to make them into works of art"*. He saw partly shaped things like an owl, or heads – of a bull, a goat, a faun, and sometimes he left nature's shaping as it was.... *"This one I didn't even dare touch: with its nose and eye sockets dug out by the sea, it looks exactly like a 'death's head'. I have nothing to add to it."* Elsewhere Brassaï quotes him as saying that: *"...man began to make images only because he discovered them nearly formed around him, already within reach"*. They were in the stones and Picasso also: *"...saw them in a bone, in the bumps of a cave, in a piece of wood. One form suggested a woman to him, another a buffalo, still another the head of a monster"*. The physical and psychic worlds inhabited by Picasso provided seeds for his imagination. And it could be argued that his full cosmos-view was used in the generation of his 'seeds'.

Limited shadowy hints like this about the nature and functionality of genius are hidden in what we've seen of human artistic and scientific creativity, and we could consider using some of these for machines. For example, the curiosity through 'surprise' of Hokusai and Picasso could be emulated in a robot, being defined in that digital domain: *"...neither as a pressure to minimize errors in prediction, nor as a tendency to focus on the most 'surprising' situations, but on the contrary as a drive that pushes the agent to lose interest in both predictable and unpredictable areas, to concentrate on situations that maximize learning progress"*.[11]

Robot Genius? The 'discovering robot' described earlier could be said to display curiosity at some level. Could other computer functionality needed or desirable for creativity be implemented in a similar way? To do so we have to be able to isolate what it is to understand and evaluate creative outputs of various kinds, as we tried to do earlier. Another possibility is to forget subjective observer characteristics and access meaning only in the characteristics of created output itself. Yet another possibility is to seek to somehow identify and then emulate what it is that the creators purpose or intend in their creations, as we did to some extent with Picasso and Einstein, and to tie this to what is generated within those who view, hear and read, or otherwise enjoy that work. This is not going to be easy, as we've seen. Cognitive states, affective states, assumptions and values, predispositions and aspirations, are among the qualities or capacities of both the creator and the 'end user' that are brought into play as they are exposed to creations.

Originality is clearly important for genius – something that has not previously existed has to come into being. A new image, idea, theorem, object or image, musical score, or solution to some open question is materialised – it becomes 'real'. There also has to be some notion of *value*, or worth of the creative output. For example, a work of art must be demonstrated to be valued when compared to other artwork. Maybe it has been shown in several galleries? Across different domains, to have been presented in some scholarly research journal, having been reviewed by several peers, to have been performed in concerts by symphony orchestras, or have been regarded as having some usefulness as a solution to a sticky problem, would do this job. As Emily demonstrated, acceptable quality and great creativity may not be all that far apart, but the sorts of curiosity, novelty and value I've been talking about seem to require some unique characteristics of human creators.

However the ability of an artist such as Picasso permeates right through the corpus of his works of art, and perhaps this gives a way ahead? In the case of musical composition, this is exploited in Emily, and the methods may be transferable, *in gross*, to literature and even fine art. Aspects of the profile of Picasso as a person is thus coded into his works, but it would be very hard for an observer to extract this from his work. Each piece of work contributes something to the overall perception anyone has of the artist. When one of his pictures – say 'The Demoiselles' – captures a viewer's attention, he/she may choose to become engaged with it, and let the 'aesthetic appreciation experience' take over. However, although this experience is surely unique to the work, the depth of this 'user' experience might be influenced by what's been seen of his work previously, and any notes on his history, and by what is known about that domain in general – visual art in this case. The picture's history, content and structure, internal consistency, and maybe even Picasso's expressed intentions for the work all come into consideration. The observer's own openness and 'soul' will be affected by the sense data input from the picture. All this contributes to the 'value' and quality of the art, and it is exceedingly difficult to isolate it.

When one considers models of style, perhaps gleaned from a corpus by a computer program, how would a human creator in any of the disciplines we've been thinking about use them? This seems to be the approach of David Cope[12] with Emily, building: *"...a computational representation of the surface of musical works, one that captures features of rhythm, melody, harmony, and structure within its patterns"*. These representations are used 'creatively': *"...to generate new music consistent within a given style"*. But it can be asked, as we did in Chapter 2, if this is really creative, in the sense above of trying something new or making something novel. The resultant models from the input corpus could be used, but in what *conceptual space* does the generated work exist? That output might be unique, but its originality is, almost by definition, in question. As I've already said, there might be some romantic evocation of emotion a

SUPERINTELLIGENCE AND WORLD-VIEWS

'consumer' experiences when absorbing a 'truly creative' work, and there might be something related to that latent within the work itself. But this is hard to measure. The originator's intent, if available, perhaps taking the form of a direct desire to transmit or evoke such feelings, might have to be encoded in the models. There is, of course, creativity here in the sense above of 'transforming old things'.

Turning from the arts to the sciences, what sort of corpus would a 'scientific Emily' use to capture the equivalents of style, for example? Could an Einstein emulator extract the essence of Einstein's work, or even something unique about Einstein himself? Could we use these to generate Einstein-like outputs to match the Emily Mozart-like outputs? Could we repeat the whole process for Newton, Leibniz, Kepler, as individuals, as well as, and then jumble all of this up and come up with a super-genius, using the sort of approach that took David Cope from EMI to Emily? This does not seem to be very likely to happen.

This sort of consideration is important for SIs. However, the question of volition arises as current computers lack world-views, and any desire 'from within' to create, and therefore they may be impotent as developers or creators of their own *conceptual spaces*. An SI that generates a work using style extraction or other analyses of genius outputs themselves for modelling will of necessity be constrained by the original conceptual space from which the corpus was generated, and which is inherent in it.

Still, could computers be used, in the mode of Emily say, to work only as ancillary 'slaves' in the production of great music, plays, poems or other aesthetic outputs, as well as scientific 'truth'? Virtual reality could play its part in this, and presumably the answer to the question is 'yes'. It might be better still if we could use computers to come up with the equivalent of Einstein's 'visual and muscular' efforts that dream up *intermediate steps* for human geniuses. Two obviously big problems are: (1) are the inspirations found by search only – as, say, in an exploring robot?; and (2) what mechanisms do they use, possibly and perhaps most desirably from our human point of view, in combination with us, to generate interesting hypotheses or 'seed creations'? Einstein did this, mostly in his brain/mind, for science, and Picasso did it in his responses to visual and motor stimulation in his brain, but predominately, as we've seen, via physical materials, like the stones and the broomstick above, and using physical media. Both of these great creators had a pattern of hashing and re-hashing as they moved towards their final output. There was 'setting up' and 'shooting down' in both cases, to use Roger Penrose's words.

I could (with a lot of help from others) design and build something like Watson or Emily. Could I build a genius, or a super-genius? Of course there are modules that I can incorporate that are equivalent to the current state-of-the-art techniques. The modules in Watson that make it better than me at fairly complex information recall are examples. However, classification and other machine learning, reasoning, and speed-related 'tricks' that go into them are insufficient on their own for genius. The SIs would appear to need genius-tricks, and they would appear to need world-views that are closer in 'fidelity to the absolute' than those I have! It would have to know the right questions to ask appropriate for the considerations in hand, and I, as the designer/builder of the SI, would have to arrange for it to do this. I would also have to ensure that it would have appropriate attitudes and understanding. These tasks will be considered further later chapters.

References

1. Bell, D. A., Beck, A., Miller, P., Wu, Q. X., Herrera, A., Video Mining: Learning patterns of behaviour by an intelligent image analysis system, *Proc 7th Conf on ISDA, Rio de Janiero*, 2007.
2. Lu, R., Zhang, S., *Automatic generation of computer animation: using AI for movie animation*, Springer-Verlag, 2002.
3. Einstein, A., 1946, *Autobiographical notes*, written by Einstein in 1946, and published in 1949: Recent edition, Open Court Publishing Co., 1979.
4. Norton, J. D., 'Chasing a Beam of Light: Einstein's Most Famous Thought Experiment', Department of History and Philosophy of Science, University of Pittsburgh.http://www.pitt.edu/~jdnorton/Goodies/Chasing_the_light/
5. Schaeffer, F. A., *Art and the Bible: two Essays*, Hodder and Stoughton, 1973.
6. Kant, E.(1790), *Critique of Judgement*, trans. Pluhar, W. S.,: Recent edition Indianapolis: Hackett Publishing Company, 1987.
7. Brassaï, G., Conversations with Picasso translated by Todd, J. M., University of Chicago Press. 1999.
8. Goodman, N,, *Ways Of Worldmaking*, Hackett Publishing Company.
9. Deutsch, D., *The Beginning of Infinity: explanations that transform the world*, Penguin, 2011.
10. Hadamard, J., *The Psychology of Invention in the Mathematical Field*, Princeton University Press, 1945. http://www.amazon.com/Essay-

Psychology-Invention-Mathematical-Field/dp/1406764191/?tag=braipick-20
11. Kaplan, F., Oudeyer P.-Y., Maximizing Learning Progress: An Internal Reward System for Development, in Lida, F. and Pfeifer, R. and Steels, L. and Kuniyoshi, Y., editor, *Embodied Artificial Intelligence*, pp. LNAI 3139, Springer-Verlag. 2004.
12. Eigenfeldt, A. A., Composer's Search for Creativity Within Computational Style Modeling, *Proc Int Symposium on Electronic Art*, 2015.

Chapter 6

Openness to the New and to the Old.

For genius level to be attained by artificial agents it would be important for them to be open to everything of relevance to the problems that they, and perhaps we, encounter and seek to solve. Our chosen 'big two' narratives, scientific and religious, and indeed all our world-views, have important influences on our conception of how far we can go with SIs. We've established that any agents that are going to be more advanced than humans will, as a minimal requirement, reflect on their beliefs, desires and actions, as we do, and wonder what to believe, and want, and do. To do this, they will have to be open to learning lessons from the old and to accommodating the new.

As far as we know, only humans can consciously and rationally assess and, if desired change, beliefs and desires, and there are a staggering range of these. Even focussing on beliefs it would require a giant step forward in technology for it to be possible for machines to do this. When we humans arrive at certain beliefs and desires after appropriate pondering, we call this *reflective endorsement*. Ernest Sosa[1], Professor of Natural Theology at Brown University, elucidates the qualitative distinction between this and other kinds of knowledge: *"One has reflective knowledge if one's judgment or belief manifests not only such direct response to the fact known but also understanding of its place in a wider whole that includes one's belief and knowledge of it and how these come about"*. So, new beliefs have to fit in with existing frameworks of reticulated concepts and knowledge. Even though we still respond most directly to immediate sensory stimuli, strong contradictory beliefs or evidence could change the response. For Sosa, reflective knowledge explicitly

SUPERINTELLIGENCE AND WORLD-VIEWS

requires that beliefs must be accommodated harmoniously within a cosmos-view, and this is done by placing it in a wider setting that will subsequently include its provenance – how the beliefs and knowledge come about. The payoff claimed is that: *"a direct response supplemented by such understanding would in general have a better chance of being right"*.

True beliefs reflect reality, and they must 'cohere' mutually because according to the stance that I'm simply assuming for convenience at present, there is one reality – only one absolute truth exists. Whether it is viewed from a religious, scientific, or any other perspective, truth must stay the same and make sense. Einstein once said that God may have made the laws of nature difficult to find, but nature would never be ultimately irrational, nor could it be ultimately inconsistent. If Einstein was right in those remarks, it must be possible to integrate all new correct findings coming from science with those laws, harmoniously even if lacking in completeness. The same is true for any new, correct findings coming from other domains, such as Picasso's visual art. Openness to 'new', correct insights should not bring contradictions, *per se*. But there is a *vice versa* here – this openness is a two-way street. New scientific and other insights should be open to old, correct givens, if any exist. It is clear, though, that the 'sensing equipment' we invoke prior to reflection when obtaining these insights is likely to vary from person to person, and we can't hope that either the new or the old inputs to the senses will allow *everything* about the other to be revealed. 'Local incompleteness' is only to be expected. My robot's sensor that is keyed to detecting green light for the purposes of navigation wouldn't be much use for detecting Higgs bosons!

Openness to The Old. The need for openness to old 'givens' such as spiritual narratives is dealt with in the early part of an influential book[2] by Allan Bloom, in which he focuses on the apparent paradox – the claimed 'openness' that the philosophical school of relativism offers can give rise to a significant 'closing' of minds. He noted something in the 1980s that just might be still the case today – that 'openness' had taken on the meaning of 'accepting everything', and he claimed that such openness had led, ironically, to the: *"closing of the American mind"*. His view was that almost all young people in the West at the time he wrote, eg many students entering university, had been left without a particular universal map or conceptual framework that had been fairly ubiquitous and robust in previous generations, when it was more or less generally accepted and was found to be, by majority consent, satisfactory, useful and easy enough to live with. Most people had taken it for granted. And yet it had simply been largely dumped, and had not been replaced in any conscious or considered way. It had not been systematically revised or updated, or consciously dovetailed with alternative views to the extent possible. A conceptual dimension was simply ignored – strangely enough because people were trying to be more open!

Maybe this particular wholesale junking was a good thing, though? Or maybe not. Bloom did not think that it was. Maybe at least the good bits could have been kept? Even some less than perfect, shared reference point whereby students and others can orientate themselves, and be consistent individually and collectively, would surely be of value to them. As long as it is correct in its main features, and as long as potential users can find out where to get one. Despite the difficulties encountered in adherence to its standards, Bloom made the strong assertion that, without his particular 'lighthouse', the Bible, which underpinned ubiquitous traditions and culture in earlier times, read with an open mind, people's minds will be 'unfurnished'. Bloom's observations of his students had led him to see something that he had not (previously) been concerned with very much. In the past students had arrived at university with quite a bit of shared, roughly agreed knowledge based on the Bible already on board and assumed to be true. This united people of all circumstances in having as an agreed blueprint for a 'standard' ordering everything of importance to them.

It is not long since there was an equally commonly consented to allegiance to a similar single broadly agreed set of ultimate truths, standards and values stretching far beyond the shores of America. Much of Western society as a whole had this dominant non-materialist outlook, FAPP, shared with many in earlier generations. Poor and rich were united, as were, for example, people from the political Left and Right, and from science and humanities, as they accepted, even if occasionally reluctantly, many 'givens', including strong values and ways of doing things. Those concepts did not solve all problems as Freud might have hoped, but they were arguably fit for purpose for a number of reasons – such as being workably consistent and clear, as well as being expressed in a fairly accessible way. Many of today's young people have or had fairly close ancestors who were often comparatively uneducated, and they frequently held only relatively modest jobs, often located in the countryside. However, somewhat contradicting the current received wisdom, those ancestors' minds were if anything *much more open* in a way that we'll see just below. And hence, they seem to have tended to be 'comfortable in their own skins'. Bloom would probably say that their home lives were enhanced spiritually as the Bible's commandments and its narratives were well known and widely applied or learnt from to good effect. Strange mental pictures and patterns of thought beyond what were necessary for daily chores and practical existence had been developed as acquaintance was made with various heroes and their adventures, over several years. Their imaginations were exercised as they were exposed to a faraway culture and its history, and their imaginations were stretched as they contemplated the infinite. They also managed to achieve much in pragmatic senses, as we know through history, in addition to living lives that were rich 'spiritually'.

Bloom noted that, by the time he wrote in the 1980's, things had changed. A different universal attitude had set in with respect to the givenness of certain 'truths'. People in general now believed, or claimed to believe, the proposition: 'truth is relative'. This belief can appear in many different forms. 'Each to their own'. 'Live and let live.' 'Anything goes (almost!)'. 'If that's OK with you, it's OK with me'. Even in matters concerning the Big Questions, and attitudes to things like ultimate truth and absolute morality, there was openness-creep that, paradoxically, was not always truly progressive. The backgrounds of those people were diverse, and yet that relativism was commonly considered to be, not a theoretical insight only, but also a moral, politically correct imperative – a must for a free society. If this has changed much since, the positive attitudes to relativism have probably hardened.

However, there is a glaring inconsistency that is implied by this, and one of the functions of the old cosmos-view was to avoid that. The coherency of relativism is very suspect. If I make a statement: 'there is no truth', I hit a well-known kind of self-referential inconsistency. What is the truth-value of that proposition? And what about the propositions that define or characterise relativism itself – are they true or relatively true? Can we say that every truth is relative except for this distinguished one – ie the one that says that all truths are relative? To be consistent, relativism itself then has to be false from some standpoints. You can't have your cake and eat it; you can't show favouritism and apply principles you won't let others use. In the absence of further evidence, the degree of falsity of a statement can be assessed rationally as being just as great as its degree of truth. One possible response to this is that a statement is true from one standpoint but false from another. The assertion 'it is 9am' could be true in Rio just now, but false somewhere else.

Restricting it to relatively mechanical and well-behaved matters like this could possibly save relativism from the charge of being straightforwardly and utterly absurd. But if we widen it out to cover statements on every kind of matter, the acceptance of multiple incompatible alternatives across the board looks like a recipe for disaster FAPP as well as logically. A well-defined commitment to truth and honesty is a basic building block of successful communication.

Truth-relativism is just one form of relativism. Others involve contextualising to, for example, historical eras, scientific stances towards the Standard Model of this or that science sub-discipline, languages, cultures and religious dogmas. Some are more significant for day-to-day life than others. Contexts, such as cultural, moral and cognitive frames of reference do not exclude one another, of course, and in fact they are usually interwoven. To take a simple example, in some communities it is a decidedly black mark to be known to drink alcohol at home, and in extreme cases, to drink it at all might be generally deemed 'bad'. Yet in France for example, the inclusion of wine with meals is ubiquitous and completely socially acceptable. The black mark there might be for someone who does not provide wine with meals.

In ultimate, extreme relativism, where we take the standpoint of the individual, we have *subjectivism*. This could potentially confine each person by the 'mind-forg'd manacles'[3] of his or her own ego. Far from leading to great intellectual and moral liberation, as many claim, the restricted intellectual space could tend to imprison people by ultimately steering them into isolation with respect to beliefs and commitments. So it is not obvious why anyone would prefer relativism to alternative schools of thought.

Among the 'open' people of previous generations that Bloom wrote about, the reason for the common consent was that their lives were founded on an accepted set of deep-seated, widely shared convictions and imperatives, in a cosmos-view that was shared FAPP. These details were, as I've said, mostly of a religious nature, at least in origin, but in practice they were by convention overwhelmingly adhered to by religious and pagan alike. They extended beyond or alongside the ritual, prayer, and narratives associated with Bible-based worship, and included strong moral postulates, values and commitments to behaviour patterns, often transmitted between parents and children, teachers and pupils, in narrative form. Some arose from explanations of those 'inherited' narratives and commentaries thereon. Reasons were readily available, if required, free of

SUPERINTELLIGENCE AND WORLD-VIEWS

charge or much effort, and free of development cost or delay, for the peoples' existence, patterns of behaviour, and their justifications for living, and for the fulfilment of their duties. They could interpret challenges they faced in the light of a line they saw running ultimately through several millennia. This ready-made, simple cosmos-view, with its associated practices and norms, provided a link from people in that generation to earlier generations who had had the same basic cosmos-view. Moreover, the people from the far past within the received narratives lived in circumstances quite different from their own, but they shared the same basic human condition and perplexities. Later thinkers and scholars at different points in history sought deep insights and offered guidance that was often timeless. The good folk at the time of Bloom's grandparents were presented with skeletal frameworks for their thinking, which featured co-ordinates, reference points and grid-lines for both received guidance and the accumulation of new intellectual furnishings. Very early Bible believers interpreted it as saying things such as that the Earth is fixed, but the enlightenment of Science, through people like Galileo, clarified the intended meaning – that the Earth is stable. Adjustments and refinements like this were gracefully accommodated as required. Einstein and Picasso were brought up within traditions from which they acquired early cosmos-views, with strong religious content, but not necessarily as commonly held as those Bloom wrote of. They were widely accepted in their respective communities as being well founded, and one gets the impression that neither of them seems to have completely dumped their received views wholesale. Old maps are not always bad *per se*. For example, having an imperfect but coherently roughed out and shared layout of one's location is better than groping around in the dark.

Openness to The New. Do we need openness to the new to accompany and counter-balance that desirable openness to the old? In one of his books[4], David Deutsch indicates that his answer to this would be 'yes'. He helps us to create in our mind's eyes, an imaginary trip back in time to the complex 'stone-age' civilisation that was evidently successful for many years at Easter Island in the Pacific Ocean. He used this illustration to help explain why openness to the new in the form of flexible learning and planning is essential for the longer-term, perhaps indefinite, sustainability of a group – in this case a whole civilisation. This island is remote, lying about 2,000 miles from Chile, and it was settled the middle of the first millennium AD. The Islanders, the Rapa Nui people, after the impressive feat of actually getting to the island in the first place, thrived well for some time, and produced a complex culture, as is evidenced by various artefacts found there when Europeans arrived in the eighteenth century. However, significantly, the population of the island had suffered a huge reduction in size from its maximal occupancy by then.

The inhabitants had built nearly 900 huge stone statues, called *moai*. The moai were big, standing about 13 feet high on average and weighing about 14 tons each, so moving them around the island was difficult. They were carved from volcanic rock quarried at a spot called *Rano Raraku*, and they were somehow transported to various locations. At their final sites the moai may have been used in some ancestor-worship-based religion, and they probably doubled up as status symbols. Nearly 400 moai in various states of completeness still remained at the quarry. According to Deutsch, following an interpretation by Jacob Bronowski, one way of looking at these statues is that they are monuments to a failing of the inhabitants of Easter Island. Under this interpretation, the Islanders were *too static* in their collective outlook. Their self-sufficiency may have been taken for granted, perhaps because life was relatively easy for them. But through time their culture disappeared as the population was decimated. That culture served the Islanders well for a long time, but their ample resources may have made them somewhat complacent, and static. As a challenging catastrophe, which was perhaps environmental in nature, loomed, they took a wrong track, Deutsch claims. He argues that the fact that a large number of statues weren't actually in place at their intended final sites indicates a serious crisis having overtaken the Rapa Nui. Roads for their transportation had been specially set up, but any environmental negatives were multiplied in the environment as the roads were made of trees. Their way of facing the problems seems to have been somewhat wishful – perhaps to bring their ancestors in to help in the face of threats. But it turned out that this was not a good way of dealing with the looming catastrophe: *"...as disaster loomed, the islanders diverted ever more effort into making ever bigger (but very rarely better) monuments to their ancestors"*. Deutsch sees the primary cause of the demise of the Islanders' culture as having been the fact that they were a 'static society'. Advance in useful knowledge became very slow or it stopped altogether, he claims. Understanding was stifled. He puts his interpretation of the cause very starkly, saying that their impressive monuments were not only a symptom of the Islanders' predicament, but they exacerbated it.

SUPERINTELLIGENCE AND WORLD-VIEWS

About a millennium before the first Easter Islanders are believed to have arrived on their island, in the first millennium BC, things were significantly different in China and numerous other places, not least Greece. Now it has to be acknowledged that we're not exactly comparing like with like here, even ignoring the big difference in dates, and some of the Islanders' tardiness can be accounted for by isolation and by the idyllic nature of life in Polynesia. Moreover, China does things on a much bigger scale to that found the South Sea islands. Certainly the land mass is very different. However, as it is attitudes and outlooks and world-views we're interested in here, hopefully those differences will not detract too much from the point being made. We could focus on a smaller region if this was considered important. Now, not unlike the case of the Islanders, the culture in China around that time was built upon traditions and rituals, and the understanding and observance of these was just as important as it was on Easter Island. However a period characterised by increased self-examination and open reflection seems to have been kicked off by exceptional thinkers, leaders and prophets. There were ups and downs of course, but overall many people began to recognise that individuals could and should actively seek, and personally reflect on new ideas and values, and that their projects should be carefully planned out and thought through. Apparently thinkers moved from place to place among the people and discussed things amongst themselves, resulting in ever better ideas. Importantly, many leaders of the day were influenced by such reflective folk. This pattern developed throughout what Karl Jaspers[5] called the Axial (ie 'pivotal') age, running for some centuries until about 200 BC, as it was in other parts of the world, such as the Eastern Mediterranean, at the same time. Yet in the earlier part of the Axial period, the society was still in a static state. But in the middle of the Axial period, Confucianism, which is claimed by many to have had a singularly enduring influence on the Chinese people, entered the scene[6]. Among the detailed specifications for behaviour that Confucianism brought, there was an important pull on people to keep learning and be open to new knowledge, ideas and skills. By widespread consent, seminal discoveries were made, and incidentally many scholars believe that in several cases this happened in China centuries before rediscovery in the West. Progress influenced social values and norms profoundly, and it was reflected in practical matters and in matters of state, where a comprehensive, thoughtful approach to the problems of government encouraged and made room for progress and achievement in other fields. Although building of The Great Wall started in the Qin

Dynasty, the Axial people appear not to have habitually made the Islanders' mistake of putting increasing efforts into almost-exclusive projects that did not address problems that arose. They did not simply increase the output of monuments to meet crises and other challenges that were arising, as the Easter Islanders seem to have done. The result was that they were well placed and appropriately trained when they needed to find ways to circumvent downturns.

Incidentally, Deutsch preaches a gospel of 'salvation via scientific inquiry', which he says is applicable even for us today. The solutions to all our problems are straightforward, given time, and (hopefully) all will be well, he claims, without establishing his grounds very soundly or even scientifically, provided we are more open to change and increasing knowledge, and more reasoning in future, so that we get 'better explanations'. Deutsch seems to assume that survival of the race, possibly through looking energetically and in an open manner for better and better explanations, is a clear and worthy goal. However, history and personal experience show over and over again that more than good explanations are needed, even for survival. Understanding how the laws of science link things is of critical importance, but the intervention of human free will has enormous impact on what transpires.

Deutsch emphasises that man is an explanation-seeking animal, but he rules out a dimension or flavour of the view that some of our case study cosmos-view holders and many of the axial people had... viz, a clear, and intellectually and emotionally satisfying, goal *beyond survival*. In many cases there was a transcendental dimension, like the one Bloom pointed out. Lacking a higher goal from 'outside the system' is not a satisfactory state of affairs for many people. As Wittgenstein and others have said: *"the sense of the world must lie outside the world"*. So they argue that we must be informed and instructed *from the outside* in how interpret our lives. The UK's former Chief Rabbi, Jonathan Sacks[7], uses the analogy of a game of football. There are rules of the game, which could be explained to a person seeing a game for the first time, and there is a corresponding internal character to the game. However the human passions aroused within the world of football, such as the intense rivalry and the single-minded pursuit of skills, can't be fully appreciated without awareness and appreciation of its wider human context.

Omega Point Cosmologies. Let's now look a little more closely at one of the 'most open' current views of the future for a moment. We've touched on it before, as it is very relevant to our discussion of

SUPERINTELLIGENCE AND WORLD-VIEWS

views on the future of science and technology, and particularly on the place of SIs within them, and, very significantly, it relates to both our two great narratives. According to the theologian and palaeontologist Teilhard de Chardin, evolution from inert matter to *Homo sapiens* is a process of 'psychic concentration' which has led – somehow – to reflective consciousness as displayed in humans, and this marked the beginning, the Noogenesis, of a new sphere of thought, the Noosphere. He agrees with many others that Geogenesis (beginning of Earth) and Biogenesis (beginning of life) come before Noogenesis, the emergence of mind. This is followed in Teilhard's chronology by Christogenesis, when Jesus Christ appeared. The current stage of evolution[8] involves the collective consciousness of humanity, and he was increasingly convinced, he reported, that the final stage will be reached through co-reflexion of intelligent beings. This sequence underpins his link-up of those two narratives, which clearly makes it of relevance to our subjects of study in this book. He sees the direction of the whole of the universe's history as being upward towards more consciousness, more complexity and more personality, and his claim is that science backs this up. But ultimately, he says, man's story is being drawn forward by a supreme manifestation of personality, complexity and consciousness, which he identified with his Omega Point. World-history will end at that distant future time…at that Omega Point. His position [9] is a *sort of* natural theology – that 'the school of science' leads to a 'yearning' that is satisfied by Jesus Christ. For some people a basic framework for cosmos-views might be derived from Teilhard's narrative and some aspects of Standard Models in physics, say, although there would be variations in the details from person to person. However, for others insights from nature are secondary and should be interpreted from the primary perspective of non-natural notions of God and his revelations.

Clearly we have to recognise our limits as future-tellers, but let's see how far we can get in assessing the value of some aspects of these frameworks. It is interesting and instructive to see how people like Teilhard and the physicist Frank Tipler link up scientific takes on The End with those of theologians and other religious believers. Information is clearly of fundamental significance to humans, and we'll see later that it is also of great significance in scientific worldviews – which are a large contributor to most modern cosmological outlooks. So we can readily imagine the development of a massive information processing facility that can help us to handle the future. David Deutsch and others calculate that taking into account quantum mechanics theory, Einstein's field equations, and related physics, it can be expected that the algorithms that may be needed to ensure human persistence in practice will require more steps than the estimated number of atoms in the universe. Those algorithms can be reached but not necessarily grasped – not yet anyhow! I'll call these *long algorithms*, as they would use spectacularly large numbers of machine code instructions. Now Deutsch has said that he adheres to an increasingly widespread, but rather strange, belief that one way to get a handle on the complex and already somewhat weird equations of quantum mechanics is to assume the existence of multiple parallel universes. According to Deutsch, Hugh Everett, and others, all events that can happen must actually happen, although, for example, my representative in any 'branch' is only aware of what's going on in that branch. Everett like Deutsch was supervised by John Archibald Wheeler (who allegedly contributed ideas to Richard Feynman's Nobel Prize), and in 1957 he introduced the idea of multiverses and claimed that all possible outcomes of observations including measurements, are actually realised! Now it should be pointed out that not all physicists are of this view. John Bell, the Irish physicist who was awarded the Paul Dirac Medal and Prize (1988) and nominated for the Nobel Prize before his untimely death in 1990, and is best known for Bell's Theorem, stated what seems to be undeniable: *"that this multiplication of universes is extravagant..."*, and he said that the phenomena being considered were explained satisfactorily by a theory conceived by De Broglie and Bohm which gives an alternative to the Copenhagen interpretation of standard quantum mechanics[10]. But leaving that debate aside - it's somewhat beyond our scope here and beyond my expertise – let's look at the Omega Point vision of the future, and The End that's presented as being possible with Deutsch's basic interpretation.

Deutsch is one of several scientists who have pointed out[11] that a characteristic of the reality we perceive in our human condition is that there seems to be a major on-going role for some sort for information managing and knowledge creating devices, perhaps to provide immersive computer-simulated 'presence' within a 'virtual' environment – a world of artificial sensory 'experiences'. Supermachines could take on this simulation role, according to Deutsch, and under this suggestion, humans would be able to plan for a manageable future. He examines the limitations of this and concludes that: *"a universal virtual-reality generator is...able to render any physically possible environment, as well as certain hypothetical and abstract entities, to any desired accuracy"*. Basic operational details of quantum computers have been sketched out by Deutsch, in what he calls the Quantum Turing Machine (QTM) and many others have taken this forward. The QTM gives an abstract model of quantum computation and it is an analogue of the classical Turing machine, which expands the classical model by allowing super-positions and interferences of configurations in ways governed by quantum physics. As we saw in Chapter 1, according to Deutsch and Lloyd, the quantum theory of computation is important for world-views, as it must, Deutsch says, be an integral part of the cosmos-view of anyone who seeks a fundamental understanding of (physical) reality. Quantum computers tell us about connections between the laws of physics, universality, and apparently unrelated strands of explanation of what Deutsch calls 'the fabric of reality', and we can make discoveries by studying them theoretically. However, 'quantum problems', such as how to deal with quantum inputs and outputs, remain to be solved. This said, the QTM idea is attracting researchers seeking to solve such problems, and also more theoretical problems. For example, it was generalized by several researchers[11] and it is known that there are several different QTMs that have different 'power' in the sense of recognizing different programming languages. The relationship between these machines is an active area of study[12,13].

A quantum computer could *conceivably* underpin execution of our 'long algorithms' on our 'massive information processing facility', even though these algorithms are a lot longer than your 'Hello World' program in Quick BASIC, or even my word-processor's program! However[11]: *"...it has a potentially unlimited requirement for additional memory, and may run for an unlimited number of steps....."* and these physical requirements mean that: *"the computer's memory accesses would have to slow down and the net effect would again be that only a finite number of computational steps could be performed"*. In practice, real matter and energy, and storage locations are needed for this, but if these aren't available in the needed quantities in the universe, as is claimed in some accounts, the suggestion seems to be that then the computer could (somehow) draw on the resources of the multiverse and cope with all the complexity. Deutsch says[11]: *"... quantum computers can efficiently render every physically possible quantum environment, even when vast numbers of universes are interacting"*.

Acknowledged support required for Deutsch's hopeful, science-salvation view has come from Frank Tipler, the physicist and cosmologist, in a number of books, including *The Physics of*

Immortality[14]. He has looked at the ('infinite') number of computational steps that it's possible to squeeze into the finite remaining lifetime of the universe or the cosmos. He came up with the Omega-Point Theory, OPT, which links his ideas up with theologians such as Teilhard. Deutsch writes: *"The key discovery in the omega-point theory is that of a class of cosmological models in which, though the universe is finite in both space and time, the memory capacity, the number of possible computational steps and the effective energy supply are all unlimited"*. The picture presented is one of life having to migrate from Earth and eventually spread through the cosmos in order to survive. Conditions will be tough though, and to survive, human essence will be reproduced on machines, that will then be classified as 'persons'. Ultimately, in fact, the only 'life' that will be possible in the universe will be machines. A mechanism for uploading personalities to very fancy computers was conjectured by Tipler as a tentative way of linking his mathematics based reflections with, for example, the reflections of Teilhard that we've mentioned. We classify the resulting agents as a sub-class of Transhumans, which we'll consider in Chapter 13. As I understand this vision of Tipler and Deutsch, the whole universe would end up as one giant infinitely powerful, sentient, information-processing facility, based on quantum computation, at the Omega Point – at the moment of 'the Big Crunch'. So we are faced again with Steven Pinker's question that we saw in Chapter 1 concerning Seth Lloyd's ideas – what would it compute?

The scientific advances to make Deutsch's vision possible would need the on-going assessment and criticism of intelligent entities – 'the persons' with those migrated personalities of ours – but not necessarily SI systems. *"This would require the continual creation of new knowledge, which, Popperian epistemology tells us, requires the presence of rational criticism and thus of intelligent entities"*. There would, of course, be benefits if they were ungraspably smart. But they would need considerable resilience: *"Space-time singularities...are seldom tranquil places, but this one is far worse than most...Matter as we know it would not survive: all matter, and even the atoms themselves, would be wrenched apart by the gravitational shearing forces generated by the deformed space-time"*. However this very lack of tranquillity could provide computational resources[11]. *"If suitable states of particles and the gravitational field exist, then they would also provide an unlimited memory capacity, and the universe would be shrinking so fast that an infinite number of memory accesses would be feasible in a finite time before the end"*

SUPERINTELLIGENCE AND WORLD-VIEWS

– the Omega Point. At the time when he was writing, Deutsch gave an endorsement of sorts of Tipler's ideas from this: *"... given that an omega-point cosmology is (under plausible assumptions) the only type in which an infinite number of computational steps could occur, we can infer that our actual space-time must have the omega-point form"*.

This vision for us – actually our 'descendants' as we saw in Chapter 1 – is seen by some as being attractive: *"...there is every incentive for them to devote their attention to managing their resources. In doing so they are merely preparing for their own future, an open, infinite future of which they will be in full control and on which, at any particular time, they will be only just embarking"*. It's clear that physical implementation is envisaged by Tipler, and Deutsch writes[11]: *"We may hope that the intelligence at the omega point will consist of our descendants. That is to say, of our intellectual descendants, since our present physical forms could not survive near the omega point."* So, this leaves room for Transhumans. *"At some stage human beings would have to transfer the computer programs that are their minds into more robust hardware ...intelligence will have to spread all over the universe in time to make the first necessary adjustments."*. It's technology – but not as we know it! Not yet anyhow.

There is nothing wrong with knowledgeable people taking a little time to speculate far into the future, and experimenting in thought like this. Full openness in the sense we're using here probably requires at least some exceptionally imaginative and ambitious prophets to stir up closed ideas and overly restrictive thinking. But prophecies should always be questioned deeply and I believe that those we've just been looking at are poor alternatives to some of the more traditional but, as Bloom showed, not necessarily more closed, thinking. It does seem that there could physically be computers powerful enough to emulate all the people that ever lived. Various estimates have been obtained when people have studied the sort of computing power we'd need for the emulation of all human history, in a *sort of* virtual reality world. Nick Bostrum estimates that in simple power and energy terms, a single computer of sufficient power for the simulation of the mental history of humankind many times over is indeed conceivable. However, power is not everything, as we'll see, and as a researcher in computers I see no reason to be confident that we will ever be perfectly copied on to computers. Incidentally, the existence of SIs as we moved towards the OPT might raise added problems – for example, would a $W^*\infty$ want a person like me who lived in 2016 to be emulated?

Tipler's claims do depend heavily on some assumptions and capabilities that it's hard to see being met even in the far future. We are a long way away from even being able to control the weather on our little planet, and, anyhow, would there be enough sufficiently detailed information actually available in practice to do the required emulation? Even if resolution of the exceedingly challenging technical issues proved possible, there are other problems looming. For example, there's the problem of getting the appropriate levels of co-operation needed to do the job. This hasn't been a strong suit for humans in history. Who could be genuinely and realistically confident that we can advance sufficiently in the future and do much better? And as I mentioned earlier, the presence of corrupt humans and the vagaries of their vested interests adds a huge dollop of uncertainty on all of these conjectures. Furthermore, unsurprisingly Tipler's link-up of the OPT with complex religious concepts is very controversial. On one hand his conjectured link-up of the narratives is not well regarded, by, for example, Deutsch, who was, as we've seen, otherwise supportive of Tipler's model. Perhaps it's *too* open for him? Deutsch says that most scientists and philosophers 'unfortunately' reject the OPT because Tipler made 'exaggerated claims for his theory'! On the other hand the match with Christian concepts that Tipler has claimed is very suspect. Narratives that say that humanity is destined for an Omega Point future depart dramatically from, for example, the orthodox Christian account of the end as being under full control of God in a way that does not match well the image of our suggested future as Transhumans. Revealingly, though, Tipler himself has said: "*I do not even believe in the Omega Point...the only evidence in its favour at the moment is theoretical beauty, for there is as yet no confirming experimental evidence for it*".

At the end of this somewhat free-thinking chapter, it's appropriate to insert a further basic word of caution here on the 'how to's' for any visions that may call on very, very good engineers to work towards as we move towards the End. *The Six Million Dollar Man* TV show was popular during the 1970s, and it encodes a warning message. After being severely injured in a crash, the hero, Steve Austin, was 'rebuilt' and enhanced surgically using new 'bionic' limbs and implants at the bargain-basement price by today's standards of six million dollars. The opening catch phrase, used by a character called Oscar Goldman, sticks in my memory: "*We can rebuild him...we have the technology*". But, I have heard a response to this in the Ulster vernacular, 'do we have the nuts and bolts and things?' These are still proving very hard to obtain. That 'bionic man', which was 'reached' forty-odd years ago in that show, and earlier, has not yet been 'grasped'. Maybe that same mirage effect remains true for the machines that Tipler predicts will be forth coming? The basic parts list is not easy to accumulate, and assembling it will be hard as well. So maybe even the most open and optimistic prophet of future technology will be forced to acknowledge the need for 'outside help' of some kind if we're ever going to enjoy existence post-Earth?

References

1. Sosa. E., Knowledge and Intellectual Virtue, in *Sosa's Knowledge in Perspective: Selected Essays in Epistemology* Cambridge University Press, 1991. As referenced by Kornblith in http://philosophy.rutgers.edu/joomlatools-files/docman-files/Kornblith.pdf
2. Bloom, A., *The Closing of the American Mind*, New York, Simon & Schuster, 1987.
3. Blake, W., London 1792. http://www.poetryfoundation.org/poem/172929
4. Deutsch, D., *The Beginning of Infinity*, Allen Lane, 2012; Penguin Books Viking, 2011
5. Jaspers, K., *The Origin and Goal of History*, trans. Michael Bullock, Yale University Press, 1953; also *Vom Ursprung und Ziel der Geschichte*, Artemis, 1949.
6. Armstrong, K., *The Great Transformation: The Beginnings of our Religious Traditions*. NY: Knopf. 2006.
7. Sacks, J., *The Great Partnership*, Hodder and Stoughton, London, 2011.
8. Teilhard de Chardin, *The Phenomenon of Man* (English version) Harper Torchbooks, The Cloister Library, Harper & Row, Publishers, 1961.
9. Teilhard de Chardin, *Science and Christianity*, (translated Rene Hague), Harper and Row, 1921.
10. Bell, J. S., Against 'measurement', *Physics World*, August 1990. See http://duende.uoregon.edu/~hsu/blogfiles/bell.pdf
11. Deutsch, D., *The Fabric of Reality*, Penguin, 1997.
12. Perdrix, S., Jorrand, P., Classically-controlled quantum computation, *Math. Structures Comput. Sci.* 16, 2007.
13. Shang, Y, Lu Xc., Lu, R., Computing power of Turing machines in the framework of unsharp quantum logic, Theoretical Computer Science 598, 2015.
14. Tipler, F, *The Physics of Immortality: Modern Cosmology and the Resurrection of the Dead*, Doubleday, 1994.

Chapter 7

Openness to the Supernatural.

If some Singularitarians are right, it is conceivable ('reachable') that our future SIs could become designers and creators of certain kinds of extensive worlds themselves. They say that it is even conceivable that an intelligent agent somewhere has already moved on to SI and become 'a kind of deity'. If some really wide thinking and open-minded enthusiasts for upward mobility are correct, some artilects might even have designed and created our universe! So religious groups have already started to sit up and take notice. Obvious problems are raised when the possibility is contemplated of artefacts that may somehow sit between humans and their creator, and what sorts of responsibility an SI would have for its beliefs and behaviour.

Religious groups may have an interest in the possible link-up with ideas suggested by Frank Tipler. As we've seen, Tipler, despite his reservations about his ideas as expressed at the end of Chapter 6, says that his OPT paradigm could provide some rational underpinning for selected traditional Christian theological concepts. He has linked his ideas with such things as the enigmatic attributes of God, like omniscience and omnipresence. Although the problem of matching terms of traditional theology with those in Tipler's framework is non-trivial, he has made a valiant effort to use it to 'explain', for example, the incarnation and resurrection. He suggests that computer emulation – roughly speaking congruent and precise copying – away in the future, will enable concepts such as resurrection and eternal life to materialise. But using one's representation as a surrogate raises questions about what is meant by 'identity'. In Tipler's brave new world, it seems to be a little bit like Susan Greenfield's 'identity' as described in Chapter 3 as 'subjective mind states'. He defines it as 'having my intrinsic properties', rather than as 'being one and the same as me', that I for one would prefer.

Incidentally, I think most people would agree with me that if they're to experience resurrection, they want it to be as themselves, in 'a body'. It is not clear that any agent would be afforded any particular comfort by the promise of the long-continuing existence of a mere representation of it. In particular, given the conditions that will obtain in the universe when the representatives are being developed, with fundamental particles being shredded and dogs yelping, most humans are, I believe, unlikely to be able to find conditions for the fulfilment of their ambitions, hopes and plans. To parody Woody Allen's quotation about death – 'I gladly anticipate resurrection, but I do want to be there when it happens!'. If the Deutsch-Tipler scenario were ever to transpire and even if I could be there as me, I could do without of all the noise and pushing and pulling of scary forces! Similarly I think most people would agree that the revealed account of a creator who wants a relationship with us in his great plan is preferable to any jumped-up designer-artilect.

These thoughts on issues that relate to SIs and at least some monotheistic versions of human religion put the spotlight on what is taken by many people to be *the main* distinguishing characteristic of different human cosmos-views. Human beings are recognised almost universally as being essentially spiritual entities, and the worship of gods is something that has been there since the advent of *homo sapiens*. It's certainly as old as the most ancient cave paintings. Finding purpose and value in life in the face of a precarious existence, and having a sense of wonder in an awesome world, seem to have always been there in the experience of humans. In the Bible, King Solomon wrote[1], several hundred years BC, that God has: "*set eternity in the human heart*" and the great English poet, William Wordsworth[2], refers to: "*A presence that disturbs me with the joy of elevated thoughts; a sense sublime…*".This could conceivably have been due partly to a felt need to be answerable to frightening superhuman but natural forces, but it cannot be denied that experiencing awe is part of the human condition, even if is often latent. This is why attitudes to the supernatural are key bases for separating out world-views of humans. A favourite way to divide world-views and cosmos-views up is to do it in accordance with how they regard the supernatural – which is closely related to one of our world-view 'flavours'. We can't do this proper justice here, but it is important take a very brief look more closely at this widely-testified-to dimension of contemplated reality.

DAVID BELL

The Spiritual World and the Material World. So the worship of gods is a phenomenon that is ancient and universal. Walter Stace[3], an influential writer on mysticism, claimed that mystical experiences are wide-spread and perennial. They involve characteristics such as a deep sense of 'the real' and of the unity of all things, of blessedness, of holiness, of ineffability, and of the laws of logic being superseded, and *"in all cultures, ages, religions and civilisations"* there seems to be a tendency to express wonder in the face of a fearful, fascinating and sometimes apparently fickle cosmos. The apprehension is frequently of a 'Universal Self', as opposed to a personal God who is a 'sovereign controller'. In many familiar religious traditions it is, however, seen as a need to relate correctly to, or at least to appease or propitiate a personal sovereign, who could, of course, be the 'something more important than [us]' that Dennett refers to.

In many traditional cosmologies, gods are said to be the origin of all living species and in some cases there is no separation between the physical and spiritual worlds – they are continuous and inter-related. A term used in the literature to cover many aspects or manifestations of this spiritual world is *vital energy*, but locally it can be *sami* or *qi*, for example, and the spiritual world can inform indigenous values and practices[4]. The sources of vital energy are variously argued to be psychological or spiritual. In the latter case its apprehension is claimed to be largely from revelation and from being connected to some higher reality, and this could be facilitated by prayer and meditation, for example, and it could inspire worship, obedience and service. On the other hand there are those who say that it can be accessed through standard natural means such as physical exercise, and in some quarters the search for an entity that fits the bill of being 'something that's not dependent on anything else for its existence'– is on-going. For example, some people take a purely scientific view and say that the universe itself will suffice. The opposing view is that the existence of one sovereign, personal God, is *necessary* – there has to be something that's independent of the finite physical cosmos for his existence. Some such 'believers' would say that God *is* separate from the physical world – 'wholly other' in a theological sense. What is clear is that reflection on theological matters touches generally and profoundly upon people's outlooks on things in general – their world-views and cosmos-views. It feeds into each individual's conceptual orientation and practical living. And of course the same might be said the other way round.

The history of the dichotomy between that view, that there is a spiritual world which antecedes everything else, and the alternative view that matter is primary, is also ancient. It certainly existed before any thought of the OPT religious issues aired just above. The 20th century saw the split becoming very pronounced, however, and this had a big impact on old certainties, as was noted by Bloom. An interesting observation was made by a character in a play by the Irish playwright in Victorian times, George Bernard Shaw[5], expressing the influence of increasingly materialistic views at the end of the 19th century. *"It's a very queer world...It used to be so straightforward and simple; and now nobody seems to think and feel as they ought. Nothing has been right since that speech that Professor Tyndall made at Belfast"*. This quote is very apposite in the present context. It appeared in a paper by Bernard Lightman[6] in reference to John Tyndall's Belfast famous 'Address at the British Association for the Advancement of Science' in 1874, where Tyndall, Professor of Physics at the Royal Institute, described his own position as a: *"higher materialism"* that found in matter: *"the promise and potency of all terrestrial life"*. Tyndall acknowledged that religion added: *"inward completeness and dignity to man,"* but he saw it as being restricted to the: *"region of poetry and emotion"*. This and related thinking at the time encouraged a thorough questioning of 'givens' about the role of religion in a modern, industrialized world and the place of mankind as a species in nature. This is just one example of the split-point between the classes of cosmos-views in which our two prominent narratives respectively dominated. In the opposite direction, Keith Ward claims that 'givens' are still given and that[7]:*"...both the atheist and the theist participate in a common faith. They both believe that reality is intelligible and that truth is worth seeking"*. Belief in the existence of God adds something, though. It: *"...provides an adequate justification of this belief, as well as an answer to the question of why the universe is intelligible at all"*. So while most people seem to understand in their heart of hearts that the material world exists, the question raised by our dichotomy is that of whether anything else exists. Both sides make extensive claims of support for their preferences, but here I simply want to look at some prominent views of what is 'beyond the material world'.

Something bigger than us. I suggested in Chapter 1 that we could give an artificial agent a particular fixed cosmos-view – to be chosen by the designer – or we could prime it, seed it and let it figure out most of its own world-views, and this represents the way to true freedom for the agent. Either option would ultimately mean the agent ending up with, among other things, purpose, values, and *perhaps* 'something more important' – a world-view Ingredient x

– to exist for. Would this be acceptable to all of those who have cosmos-views which include some of the standard spiritual beliefs? Taking the first option above and possibly avoiding the most difficult components of this issue by arranging for the key world-views of a very smart machine to be given to it in detail and controlled by humans, would make the classification of the artificial system as 'an SI' rather suspect, as I suggested in Chapter 2. To be truly SI the system would have to understand the cosmos-views of humans, and maybe have a cosmos-view of its own choosing that fits in with the more credible varieties of those human views of the cosmos. Looking at the 'future telling' aspects of human world-views also helps us to keep a good perspective on any developments 'towards SI'. So we need to see what options there are for humans to choose from with regard to a spiritual 'something else', and consider how SIs might fit in with these. For these reasons in this Chapter I sketch a backdrop that will hopefully be useful in our thinking about these things.

The set of graspable things is, in the sense I'm using, a strict sub-set of the things that are reachable, and so, of course, is the set of fully believable things. Maybe that's all a bit too far out for us. But we can readily think of 'more scientific' things that we can probably reach but not always grasp – like quantum gravity, quantum logic or the Higgs boson, or like quasars, or simply the square root of *minus1*. The well-known applied-versus-pure tension in some branches of science or engineering research where too much focus on the graspable or 'applicable' at the expense of the reachable, or 'disinterested', could limit exploratory activities. On the other hand, over-focussing on reachable but not graspable things can arguably desensitise us to, and blunt our appreciation of, other highly important aspects of existence, and getting this wrong can leave us dreaming, bedazzled and making little progress FAPP. Maybe our all-consuming focus could even be on some great other worldly cosmos-view, and then, of course, there is always a question lurking in the back of our minds: Is it all an illusion? Umberto Eco, in *Foucault's Pendulum* raises the question[8]: *"But if there is no cosmic Plan? What a mockery, to live in exile when no one sent you there. Exile from a place, moreover, that does not exist."*

Faced with these perplexities, and the uncertainties that are built-in to our existence as acknowledged by modern science and mathematics, and we'll look at some of those in Chapter 12, we try to negotiate the bewildering mishmash of curiosities, apparent inconsistencies and even ostensible absurdities in which the 'jigsaw pieces' do not quite fit together, and where appearances may be very misleading. As I've said, humans seem to have always looked for explanations of their world – in particular the world of nature as it impacts on day-to-day existence. When faced by some deep questions arising in their lives, such as the Big Questions we've mentioned before, they often simply accepted the authoritative explanations given by those that were regarded as being wiser and more knowledgeable than they may have deserved. At first explanations of basic natural processes, such as the 'spontaneous creation' of maggots in rotting food (see Chapter 12), were treated in the same way. Alternative explanations were simply deemed unimportant. Some outliers, free and original thinkers, have always existed, but dissent tended to be put down, and dissenters tended to be marginalised.

Mind or simply Matter? Where exactly does a God who is a *person*, with a mind, and the supernatural in general, fit into peoples' pictures of 'something beyond'? For example, views of the universe as a giant, impersonal Turing machine, that we're moving closer to becoming more in harmony with, do not go far enough. For example, Sir James Jeans, a mathematician, physicist and astronomer[9] is frequently quoted as claiming: *"Mind no longer appears as an accidental intruder into the realm of matter. We are beginning to suspect that we ought rather to hail mind as the creator and governor of the realm of matter...".* This is not quite the same, of course, as saying that the universe will be, *in toto*, a great mind, the full manifestation of which possibly needs to be worked towards by human artilect designers and others.

We saw earlier that Albert Einstein referred to: *"...the mind revealed in this world"*. Many other famous scientists had similar views. The idea of there being a mental aspect to the cosmos is broadly agreed with by many others. Nobel laureates, like George Wald in biology, Werner Heisenberg and Max Planck in physics, and Christian Anfinsen in chemistry, along with other prominent scientists, are among other leading thinkers who make similar statements, which are perhaps surprising to those exposed to familiar propaganda. The following is a selection of quotations from these scientists: *"It is Mind that has composed a physical universe that breeds life, and so eventually evolves creatures that know and create.... there exists an incomprehensible power or force with limitless foresight and knowledge that started the whole universe going"*.

Of course, recognition of mind as part of the basic fabric of the universe does not always mean acceptance of revelation and authority as sources of fundamental knowledge, but many scientists do go beyond this *deistic* stance. In his book *Infinite in All Directions*,

Freeman Dyson[10], another visionary theoretical physicist, who was awarded the Lorenz Medal in 1966, The Max Planck Medal in 1969, the Harvey Prize in 1977 and the Enrico Fermi Award in 1993, identifies three levels of mind. At the elementary particle level: *"Matter in quantum mechanics is...constantly making choices between alternative possibilities according to probabilistic laws."* The second level: *"...is the level of direct human experience"*....He sees it as being rational to see the operations of mind in the universe as whole and calls the mental component of the universe 'God', so that we can then say: *"...that we are small pieces of God's mental apparatus"*. He goes on[10] to declare his relationship to that great mind as a 'Christian': *"...a member of a community that preserves an ancient heritage of great literature and great music, provides help and counsel to young and old when they are in trouble, educates children in moral responsibility, and worships God in its own fashion"*. He then goes on say that: *"...to worship God means to recognize that mind and intelligence are woven into the fabric of our universe in a way that altogether surpasses our comprehension."*

The perceived nature of 'sovereign controller' varies greatly within and between the codes and traditions of various groups, and Dyson says his is a Christian one of sorts. Now, many people of a certain vintage with a Christian background have memorised written summaries of their beliefs. One question in one such written summary or catechism is: 'What is God?' and the response to be remembered is 'God is the Supreme Spirit, Who alone exists of Himself and is infinite in all perfections'. A somewhat different view, though, that's held by some theologians and others, still claiming to be under the Christian banner, is that the world is 'a riddle' that God created to shield himself from terrible terror and solitude. Some of those holding this belief follow Socinius, who in 16th century, believed that God 'learned' and this implies that he is *contingent*. According to the Abrahamic tradition and the Bible, God said, 'I will be what I will be'. On the Socinius view, God doesn't know, right here and now, all details of future events and they won't be known until they happen. In its extreme form it is not much use as an absolute fixed-point, and it is clearly very different from the catechism response above. It's presumably not what most people would want to pin their ultimate hopes, trust and commitment on. Now that's an illustration of just one such controversy even within one tradition of one faith, Christianity. The potential for confusion is clear here, and even the key distinctive feature of Christianity – the absolute necessity for faith in Jesus Christ as Saviour and Lord – may be missed.

SUPERINTELLIGENCE AND WORLD-VIEWS

The End. Let us look at some rather more perplexing, in the sense of being impossible to know or even fully hold in our consciousness, concepts that a lot of people put to the back of their minds, but which are claimed to impact very significantly on our futures. Often this (hopefully) temporary, deliberate covering can be justified FAPP, but those thoughts need to be taken out and looked at from time to time, and definite views on them should arguably be taken by every mature, thinking human being, especially those aware of and interested in cosmos-views.

Fear of annihilation is one reason, among many, for a 21st century interest in the ultimate future. There is a lot of talk these days, and in fact it's been going on at many levels for quite a while, about 'ends of eras'. For example, when a great sportsman retires, or when a government changes after a long time in power, we might use the term. But the general idea of significant termination is of much wider scope. The End of History[11], The End of Science[12], The End of Nature [13], The End of Christianity[14] and The End of Reason[15], have all been written on recently. In a similar way 'crises' are flagged up. Umberto Eco says[16]: *"During the last few decades we have witnessed the sale (on newsstands, in bookshops by subscription, door-to-door) of the crisis of religion, of Marxism, of representation, the sign, philosophy, ethics, Freudianism, presence, the subject"*. He also mentions crises that are claimed of the lira, housing, the family, institutions, and oil, respectively. The existence of this perceived tsunami of crises was confirmed for me by a brief look at the first page of the Google response to a request on 'crises'. It gives, for example: The Crisis of Reason: European Thought, 1848–1914; The Crisis of Authority in American Evangelicalism; Six Reasons Another Financial Crisis Is Inevitable, and The Hidden Reason for the Student Loan Crisis. Eco suggests that, rather than using the word 'crisis', we should perhaps be using the word 'critique', by which he means: 'the recognition of limits', and which is not quite so scary.

John Horgan[11] gives a thought provoking, if somewhat 'woo-woo', to use the term he used, look into what the prospects might be when science and all other modes of enquiry have ended. He talks about the furthest out, probably unrecognised and unacknowledged, goal of searches for knowledge as being to ultimately 'extinguish wonder'. This is, incidentally, against my own view on this. I believe that increasing illumination from new discoveries *increases* wonder at uncovered intricacies and grandeur. Horgan does, however, depict a very other -worldly climax of all things. When that is reached, he says, metaphorically speaking, the question: 'What's it all about?'

will be transformed into the 'Aha!' when The Answer is found. He suggests some other esoteric possibilities, such as the unveiling of a single divine entity embodying the 'spirit of Christ', as suggested by Teilhard, who of course called this climax 'the Omega Point'. A slightly different but related description is set out by the German Christian theologian Wolfhart Pannenberg whose belief on The End – the Eschaton – is, according to one reviewer[17]: "*When human beings enter eternal life, then, the final future which brings history to an end, they enter into God's own life, where they experience the full expanse of their historical existence in one simultaneous moment of perception*".

In one way, we have to take our hats off to people like Deutsch and Tipler who venture into the sorts of speculations we've been looking at. They imply we have to work towards a climax, which will be there at The End. But, if, as some religions teach, this is all away beyond our reach, never mind our grasp, as humans, maybe this sort of speculation is not so admirable, and it would not be wise to put huge intellectual efforts into stretching beyond what it is possible to appreciate. There is sentence in the Bible[18] which says: "*Eye hath not seen, nor ear heard, neither have entered into the heart of man, the things which God hath prepared for them that love him*". Indeed when we try to figure out what a finally fulfilling future could look like, we flounder. We are 'hit for six' by our wonder and imagination, as we contemplate, as an example, Pannenberg's views on how God's relationship to the world will play out and culminate. His 'event' is described as 'time giving way to timelessness'. Does this mean a transition from a continuous life, where we don't have everything happening at once, to one where we do? Temporal experiences, seemingly irreplaceable in finite existence as we know it, giving way to a final future labelled as 'timeless', is as hard to get one's head around as anything I, for one, can think of. God's dealings with the world are usually revealed in terms of time, so if his own existence does not include some sort of time-like, sequence-based, characteristics, then these dealings might not 'soundly and completely' reveal God after all! Indeed, taken alone, they could then be seen as devaluing and maybe even misrepresenting, God's essential nature. That's why we flounder, and thus we have a very fine example of a case where our reach exceeds our grasp. Moreover the problem of imagining what the Utopian future will look like also looms large, as we've seen earlier. Horgan[13] seems to think there might be some attraction for some in somewhat basic, rather earth-bound activities. He mentions making;

"ever-more-clever-conversation with ever-more–beautiful super-models". Is that what it's all about?

The Transcendental. Some patterns of thought have occasionally been criticised as not seeking or recognising the oneness of everything. For example consider the warning the poet Goethe gave about the danger of too much *Zergliederung*, or 'tendency to over-analyse': *"They proceed to pull apart. They demonstrate the truth of the object, until none believe it anymore"* [19]. The theoretical analysis essential to, say, *naturalism*, which excludes the supernatural, is needed for improving our understanding. However, we can have too much of a good thing. Having too many alternative explanations and predictions is confusing. Even in everyday life, keeping a simple coherent global 'oneness' perspective, through our cosmos-view, can limit the likelihood of 'analysis paralysis' – ie being faced with too many choices. However, we must keep a balance – and be wary of overstating this case, as 'reductive' understanding of nature underpins much practical progress we've seen in the last century or so.

What about the alternatives to naturalism that *do* consider the supernatural? The quick sketch I now give will be slanted a little toward my own view. So I omit many possibilities, such as Koestler's beliefs and faiths in which individuals look within for inherent purity, perhaps is through meditation in solitude, or in 'oneness with nature'. I just mention in passing however, that simply trying to understand 'the mind of the universe', or 'enlightenment' or 'unity with the divine' leaves unsolved the problems of, for example, the existence of evil or predicting The End.

Theism, adhered to by the likes of Pannenberg and Teilhard[20], which means the belief in a personal, self-revealing God, is one important alternative. Theists whose views I am familiar with believe humans are created in the image of God for relationship with God, that the source of man's problems is disobedience to God, and that the solution is to be found in God. They believe that humans know God primarily through supernatural revelation and illumination. Their analysis accommodates natural causes where appropriate, but it does not insist that all events are to be explained in this way. Natural processes themselves are often seen as being planned, created, sustained, guided, and on occasion interrupted, by God. This has a big impact on the corresponding world-views of the theists, and we've already seen that there are many variations among these. Some say we can, and maybe have to, appease God, who is the 'sovereign controller' referred to just above, for our short-comings, or that they can accumulate 'Brownie Points' by doing good things, in order to receive some sort of positive balance, in some sort of currency, with God. Others believe that man's redemption is simply a 'gracious gift' of God. Some, but by no means all, believe that God is involved, hands on, in the micro-management of the universe, and some say that he is knowable, others that he is not. *Deism* can be seen as a special case of belief in a supernatural God but it suggests that there is no need for revelation. It claims that God is beyond the world and *not involved* in it at all, and he does not intervene in it, but lets the natural world run according to laws he set up at start-up time.

Alongside Christianity and Islam, Judaism is a one of the three great monotheistic religions. Jonathan Sacks summarised his view as Chief Rabbi in the UK at the time he wrote, in a way that captures one version of the essence of the theistic stance[21]: *"The universe was called into being by One outside the universe"* and that 'One' let it develop. *"Eventually life formed and evolved, until one creature emerged capable of communication. The One sent messages to this creature...It is said that every human being had within him or her a trace of the One who created the universe...."* In the course of time religions related to Judaism, such as Christianity and Islam, appeared. Through each of these religions, Sacks claims, *"...in striving to listen to the more-than-human, human beings learned what it is to be human, for in discovering God, singular and alone, they eventually learned to respect the dignity and sanctity of the human person, singular and alone...within us is the breath of God"*. Incidentally, in his summary Sacks uses the phrase: *"The One waited to see what would happen next"*, which gives another example of how controversy soon arises between religious stances, as it quite like those of Socinius, and Sacks' strong hint at an evolutionary past of mankind would be opposed by many theists.

Revelation is important in all three Abrahamic religions. In particular, in the Christian faith, as understood and practised by, say, David Livingstone, revelation is seen as underpinning a personal and life-changing encounter with God, and as showing that everything we need to know about God is to be found in Jesus Christ. Any account of revelation has to cover the diversity within and between those three religions and others, but that is beyond our scope here. I just want to note that at base, revelation has to be concerned primarily with *absolute truth* that is being revealed. A key part of the Christian message or *kerygma*, is that what goes on in the 'phenomenal' world affects what goes on in that eternal world, and *vice versa*, although there is a presumed precedence relation between these goings-on. If this is correct, then our cosmos-views will lack some dimensions and flavours if our considerations are restricted to time, space and sense. And of course, in a parallel with Sack's analogy with football referred to in Chapter 6, God has to traverse possibly unknowable wider dimensions if his revelation is to warrant absolute confidence. But it is worth noting that even when we include the *extra-physical* in our views, we still encounter great opportunities for openness, manifested by the fact that there are more choices to make in establishing our cosmos-view, even for a relatively restricted part of the belief spectrum. In Christianity, for example, there is no revelation *at detailed level* of the future of mankind. Neither are there any details of a clear analogue of the Tipler-Deutsch shreddings.

So the two great narratives we've been running with, and the associated kinds of cosmos-view exemplified by, say, the Deutsch and Sacks pictures, respectively, might at first sight be seen as being clearly at odds. According to many scientists, though, they can be reconciled. Would this be easier if we had the superintelligent machines we've been talking about – say W*25 and beyond – to help us? Or would that make the reconciliation harder? I am raising such questions because they relate to the possibility of SIs and how the SI concept impacts on our own human world-views. Other outstanding questions have to be answered, of course. Would we have to ensure that any future W*22–equivalent is given, say, Tipler's cosmos-view? In due course we'll be looking a little more closely at some of those more universally prominent features of our human world-views and cosmos-views, with a view to assessing the likelihood of machines having them. However, important questions have clearly been raised at this point, such as: Would a super-intelligent machine need to accommodate transcendence in its cosmos-view? A related question is: can a machine be religious?

References

1. Solomon, The Bible, Ecc 3:11.
2. Browning, R., Lines Written a Few Miles Above Tintern Abbey, 1798.
3. Stace,T., *Mysticism and Philosophy*, Macmillan,1961.
4. http://www.review.mai.ac.nz/index.php/MR/article/viewFile/186/196, MAI Review 3, Research Note 5, 2008.
5. Shaw, G. B., *Man and Superman*, 1903.
6. Lightman, B., "On Tyndall's Belfast Address, 1874." BRANCH: Britain, Representation and Nineteenth-Century History, Ed. Dino

Franco Felluga. http://www.branchcollective.org/?ps_articles=bernard-lightman-on-tyndalls-belfast-address-1874.
7. Ward, K., *Why There Almost Certainly Is a God: Doubting Dawkins*, Lion, 2008.
8. Eco, U., *Foucault's Pendulum*. Weaver, William (trans.), Secker & Warburg, 1989. http://wwwPDFs/9780230365.palgrave.com/476.pdf.
9. Jeans, J., *The Mysterious universe*, Cambridge, 1931.
10. Dyson, F., "Science & Religion: No Ends in Sight". The New York Review of Books: March 28, 2002.
11. Fukuyama, F., *The End of History and the Last Man*, Free Press, 1992.
12. Horgan, J., *The End of Science*, Addison Wesley, 1996.
13. McKibben, B., *The End of Nature*, Anchor,1989.
14. Dembski, W., *The End of Christianity*, B and H Publishing, 2009.
15. Zacharias, R., *The End of Reason: A Response to the New Atheists*, Zondervan, 2008.
16. Eco, U., *Faith In Fakes: Travels In Hyperreality*, Picador 1987.
17. Rice, R. W., Pannenberg's crowning achievement: a review of his systematic theology, *Andrews University Seminary Studies*, Vol. 37, No. 1, 55-72, 1998.
18. Paul the Apostle, *The Bible*, 1 Corinthians 2:9.
19. Googlebooks: Goethe's poetische und prosaische Werke in zwei Bänden, Volume 1, Issue 1, Gedichte 131 By Johann Wolfgang von Goethe, Friedrich Wilhelm Riemer, Johann Peter Eckermann. http://www.generation-online.org/p/fp_simmel1.htm
20. Teilhard de Chardin, P., *The Phenomenon of Man*, (English version) Harper Torchbooks, The Cloister Library, Harper & Row, Publishers, 1961.
21. Sacks, J., *The Great Partnership*, Hodder and Stoughton, London, 2011.

Chapter 8

Scientific and Transcendental Views for SI?

When we look at modern and realistically-projected computing facilities, that question of whether or not a machine can be religious looks a bit ridiculous. With regard to certain key, technology-related capabilities, in hardware, software and even theoretical directions, it seems to be obvious that there are also very important lacks – features that a robot or similar artefact would need if it could be said to have its own transcendental experiences.

Let's consider for the sake of argument that it has been deemed desirable for some reason to have religious machines. There are two different approaches that designers and developers of more human-like machines and SIs could consider. To start out to build these features progressively into existing artefacts would be a bottom-up approach – designers decide where we are at present and judge what our next steps will be. An alternative is to focus on scenarios nearer 'the top', on what things might someday be like. Towards the end of this chapter I will look at the sorts of offerings and visions coming from science fiction and the like, and note a serious problem – they don't provide a viable unbroken link either with today's technology or science or with the results of philosophical and theological reflections. I will argue, moreover, that many of the current offerings from the future-reading side of, say, the entertainment world, in science fiction for example, fall well short of the required quality. So we will probably never want to get to there from here.

I'll (greatly) simplify, of course, but before that I also want to emphasise the difficulty of accommodating transcendental worldview components alongside high-quality scientific and technological component views. Hopefully what I mean by high-quality will become clear as we proceed. Many humans seem to be able to accommodate these tolerably harmoniously even if one or other has to give way a little, for use in explanation and prediction cosmos-view functions. There is a real possibility that machines may be permanently doomed to failure on this, due to its sheer complexity and subtleties. As we look at a particular Big Problem – that of the origins of humans, and incidentally the origin of morality, it is good to have some basic appreciation of what some pertinent human scientific and theological 'state-of-the-art' views are. The question of what place these components might have in the SI's own cosmos-views is also raised. As well as trying to accommodate both in some way, SIs would presumably have to understand that they could be in conflict for individual humans they interact with, and between individuals and groups.

In the next few pages I show briefly the complexities this might entail by homing in on an interesting and important corner of each of our 'two great narratives', which might be considered for inclusion in an SI's cosmos-views or simply its modelling of humans. These are things people choose to believe, or not, because neither of these narratives is provably sound and complete. This is not surprising in the case of theological world-views, but even in science we can be led astray in the received evaluations of the 'historical' science. We will see in Chapter 10 that Henry Gee and others have warned about the dangers of emotionally 'piggy-backing' on the reliability of familiar 'operational or experimental science' and carrying the level of confidence we have in our understanding of cars, cookers and mobile phones, for example, over to 'historical science'.

A Scientific Development Scheme. I am now going to give an example of that particular potential and serious error. I'm assuming here for the sake of simplicity, a working conceptual framework for the historical scientific narrative. This includes a widely-accepted *development scheme*, a *sort of* Standard Model for the early developments in the human species since it originated, for which there are multiple sources of very suggestive evidence to point to. In recent years, decades and even centuries, we've seen overwhelming growth in our understanding of the world. Impressive strides forward have been taken in for example, palaeontology, geology, cosmology, physics, and biology, and many of these have important bearings on our present topic. To set the scene, consider some photographs you can find easily on the Internet by searching with keywords like Laetoli or Afar. One photograph, dating from the 1970s is said to be of a trail of footprints about 25 metres long, of an animal that has since been labelled *Australopithecus afarensis* after the famous British palaeoanthropologist Mary Leakey's discovery in Laetoli, Tanzania. The trail is embedded in a series of overlapping surfaces that are the result of volcanic flows and falls of volcanic ash, thought to have been laid down bit by bit and quickly cemented by rainfall, from about 3.7 million years ago. The narrative that is used to describe and explain the photo's content presents a key challenge to a 'transcendental' narrative that we'll see below concerning the advent of humankind, and the origins of a key theological concept – 'The Fall of Mankind'.

The footprints in the photograph are those of an individual which/who has famously been called 'Lucy' after a Beatle's song that was popular at the time of discovery. She was essentially 'a survival machine' that walked somewhat more uprightly than other chimps. It is claimed that this evidence shows, among other things, that human feet did not instantly appear, but had an extended evolutionary history with various developments taking place at different times. The prints are rather inconclusive but they are quite human-like at a certain level of similarity. It is said that *if they were not known to be so old*, people would say they were made by humans. Because of the dates, however, the prints have been assigned that scientific binomial, *Australopithecus afarensis*. The first part of the name identifies the genus *Australopithecus* to which Lucy belongs; and the second part identifies the species within the genus, named here after the Afar Triangle in Ethiopia where other Lucy-like footprints were found by a scientist/explorer, the American palaeoanthropologist, Donald Johansson. Johansson put two and two together and linked a previous discovery of his own with Mary Leakey's discovery to come up with a judgment on the knee structure of Lucy, and conjectured the oneness of the species at the two sites. It is claimed that although Lucy was essentially an animal honed by the environment for survival she was an early human ancestor. She has been likened to a chimpanzee, seemingly being capable of climbing, for example, but it is believed that she walked more or less upright most of the time. My point here is that this sort of evidence is not of the quality one would expect to establish facts about, say, the properties of electrons, so important in our mobiles and lighting systems, or the properties of internal combustion engines for our cars.

Consider now the identification and dating of the various Lucy bits and pieces, important to the conclusion that Lucy was one of our own ancestors. Various methods of fossil dating are now fairly standard, but they are not infallible, of course. Elements such as

Carbon 14 (C^{14}) are believed to decay at a constant rate, so an estimate of the date at which something died can be found by measuring the amount of the element in a specimen. In the last few decades *potassium-argon (K-Ar) dating* has been used to suggest dates for strata as far back as about 4 billion years. For Lucy, geochronologists using it produced an estimate, 3.6 million years, of the age of the rock around her as she lay. However, although we know that K-Ar dating works well enough over time within timescales that can be verified independently, say from historical records, the established reliability of such methods does not stretch back to events 1,000 times longer ago. Furthermore, the method does have several recognised limitations[1]. For example, in potassium-bearing minerals argon can be trapped as they cool and crystallize, and all of the events involving the rock may be uncertain. So, discretion and interpretation of age dating is essential. However, armed with these dates and bolstered by some evidence from palaeontological studies, for example, researchers place the Lucy fossils into a date framework somewhat tentatively, as estimates of age are being revised all the time. This uncertainty is compounded by the sketchiness of the evidence outlined above and the significant reliance on other more basic assumptions – eg that all we've observed and recorded over the last few hundred years has been unfolding in just the same uniform way for timescales that are many thousands of times longer. Those time-scales cover very many cycles of Earth-wobbles, and frequent tilt variations and orbit shape changes, and these have to be factored in to this chronological narrative or Standard Model for the evolutionary development of our species. Asteroid collisions and other catastrophes could also have had impacts during that multi-million year window.

However this is not the only evidence we can bring to bear on the task of tracing the human story. Consider the evidence coming from another of the domains mentioned above – from the human genome projects and genomic information in general. For example, a plausible enough path can be conjectured from simple chemical elements, through polymers that become able to replicate, and a couple of other steps, to bacteria. Moreover, although sizeable human lineage-specific genomic differences between humans and primates have been known for several decades and knowledge has rapidly increased in the last decade[2], shared genetic material is pointed to as providing *evidence for the inter-relatedness of species*, and, it is of the sort one would need for some degree of micro-evolution. Now this evidence by itself is, of course, not conclusive in an 'operational

SUPERINTELLIGENCE AND WORLD-VIEWS

or experimental science' sense. For example, the sharing of characteristics *per se* does not imply common ancestry. As we saw in the case of some 'human ancestors,' temporal inputs are needed to make that record into a historical development path. Similarities between 'species' is not enough by itself for reliable validation of origin conjectures. A man-made chess player and a natural chess player could share a lot of characteristics (although there are also, of course, many dissimilarities), but their origins and sources are quite different.

Other disciplines such as geology and archeology also offer trajectories of understanding independently of biology, and each of these can be used as an additional source of evidence about an aspect of a cosmos-view. There are provisos again; for example, different dating methods are frequently used to back each other up, sometimes in a suspiciously circular manner, often to fit with the current scientific world-views. Moreover, results must be challenged in the light of discrepancies as in other branches of science. Again I emphasise the point made by that Henry Gee and others who have warned that the predictability of scientific phenomena in the lab and over small time-scales, well-confirmed in our experiences, eg with technology in the modern world, cannot be matched by non-repeatable 'scientific results' in those domains we're talking about– but also, for example, in cosmology.

In spite of these weaknesses, each independent trajectory of discovery showing improvements in understanding can be seen as a thread, where each thread can be seen as providing only weak evidence for a Standard Model if taken alone. However, even taking account of the possibility of inaccuracy, bias and some circularity of reasoning, they do weave together into a somewhat stronger cord with which to tie a lot of concepts and models together into a whole. Even if confidence in the results and findings is very low compared to operational science results, the sub-narrative is therefore reasonably self-consistent. But it must be accompanied with an appropriate dollop of strict scientific caution or even scepticism. Following a *sort of* convention adopted for this book, I'll call the rough narrative reached by such reasoning the Standard Model for Human Development. While this model is by no means assuredly correct in all respects, it does provide hard challenges to anyone seeking to completely refute it in a coherent way, so I'll use it as a very provisional, rhetorical *reference framework* for our discussion. It could also conceivably serve this purpose in the case of SI cosmos-views.

According to the timeline of this Standard Model, the balance of the evidence that's around these days is said to support the conjecture that other anatomically modern hominids preceded us, the *homo sapiens* folk. According to the Standard Model, these hominids were, taxonomically speaking, a branch of the great apes with larger, more complex brains, and with more complex societies than other apes. Furthermore, increased dexterity of hand, meat eating and tool use were also distinguishing features of these populations, and they walked upright on two legs. Alongside the earliest humans there would have been survivors of these earlier hominid species, maybe *Homo erectus, or Homo neanderthalensis*. The current consensus seems to be that we have all descended from one Lucy-like woman[3]. These individuals were, according to this narrative, parts of larger populations with whom their descendants interbred.

As widely presented, the development scheme we're talking about includes a 'migration scheme', which in the Standard Model starts out with a small population of hominids living at some spots in Africa about 100,000 years ago. Populations then spread out around the continent as a whole, and into Europe, Asia, and further afield. There is some evidence, again by itself quite weak, to support the proposition that common ancestral *homo sapiens* individuals appeared on earth much more recently. So science can be said to agree reasonably well with Sacks' statements in Chapter 7, and also with the ancient scriptural proposition that we humans are all of one species. Dates are aligning better all the time with those when God would have 'breathed the breath of life' into Adam, and he became the first human 'living soul', 'in the image of God', having impressive cognitive capability and capable of having a relationship with God.

A Theological View. Finding agreement is more difficult as we try to marry this standard narrative with all details of 'higher' world-views and narratives, which indicate that we are created in the image of God, for relationship and co-operation with him. As Karen Armstrong[4] says, and as we saw at the start of Chapter 7, the human condition is widely taken to include religious tendencies. *"When people try to find an ultimate meaning and value in human life, their minds seem to go in a certain direction,"* and this is a persisting feature of humanity. However, hard, directly supporting sense-data-based evidence is in as short supply in the case of *revealed* origins of the ultimate source of *everything* – ie not *just* humankind's origins - as in the Standard Model just considered. Incidentally, science has also had little to say on this – even less than it has had on human origins. Indeed operational science cannot even address the ultimate question of origins. Take the story that is developed around the fact that the pattern of 'quantum fluctuations' presented in the 'best' cosmological and quantum equations agrees with observations of cosmic microwave background radiation, and the pattern is claimed to have 'emerged' afterwards: *"as density fluctuations that produce everything we can see!"*[5]. The well-known USA Professor of Physics, Lawrence Krauss, says that the existence, now, of: *"galaxies, stars and planets and people"* is due to, *"quantum fluctuations in what is essentially nothing"*. This concept of 'nothing' is very reminiscent to me of what I've read of ancient pagan myths, or the Gnostic thought of Basilides in second century Greece, or the speculations of Kaballists such as Isaac Luria in the 16th century. Krauss himself admits that all this can sound: *"almost magical"* or *"a bit fishy"*. Little physical detail on this is offered from theological sources and revelation – but there is more illumination on 'whence mankind' and on concepts such as formal promises that God is 'revealed' to have set up with mankind in the earliest times. The Fall of Mankind and Original Sin, the stated inevitability of God's retributive justice and the promise of salvation by grace through faith, all lean entirely on revelation, as they are all likewise outside the sphere of competence, interest, authority, or skills of science, in spite of the best efforts of people such as Frank Tipler. Simple modelling of religious concepts in machines, for example, raises some interesting and very, very challenging problems.

Still, how the spiritual is handled remains, by common consent, a key feature when cosmos-views of humans are classified. So the designers of SIs will probably have to at least think about including it, or replacing it. As pointed out by Karen Armstrong, the fact is that many humans believe that mankind would be impoverished, and perhaps even 'calcified into non-being', if spirituality were lost. Atheists would of course say that the lack of these sorts of features would not present a problem, but it is clear that we still have some way to go to the point where we have SIs with at least the possibility of including these sorts of concepts so that we have the *full equivalent* of human cosmos-views. As we'll see in a minute, though, even science fiction writers and the like find it hard to provide a viable link between their characters and their 'experiences' and today's technology and science, and with the results of theological reflections. Let's now look a little more closely for illustration, for a few paragraphs, as an important part of this quick look at the state-of-the-art of 'SI development', at the possibilities of one of those 'higher' doctrines being communicable to the agents in the world of machines.

Suppose we take the Christian doctrine of Original Sin that I mentioned above as an example. This is my 'interesting and important corner' of this narrative. To start with, we would need a good 'high-level specification' for our designers and programmers, human or mechanical, to target, and there's no unquestioned Standard Model here either! However there is a reasonably well-accepted account of this, as set out, in the version I've chosen, by Henri Blocher, a Christian theologian I'll be referencing from time to time. He has explored [6,7] thoroughly the hard theological questions related to the origins of man and sin. We take some time here to look at this in detail, as the wider considerations concern values and are a very important aspect of cosmos-views, and it is interesting to note how smoothly it can be linked up with the cosmological Standard Model we've just talked about. Blocher claims to have a reliable source for this particular component of one widely held cosmos-view as he considers the question of there being a definite historical beginning of *sin* – at the Fall. It's the same source as Bloom referred to, ie divinely revealed scripture. Blocher also submits his view that the account of our earliest human history given in this source provides a *conceptual framework* that describes God's work in creation and events following very swiftly from the advent of mankind, which, though he says that it is not meant to be a literal procedural account in all its details, and definitely not a scientific account, of all facts, it is indeed completely *historical*. Blocher indicates that he can live with a relatively large age estimate for the human race and lose nothing of importance theologically. In fact he gives three 'Adam-times' for the Fall: Neolithic Adam or Genesis Adam about 9500 BC; Cro-Magnon Adam about 32000 BC, the suggested time of the Grotte Chauvet paintings in France; and a Carmel Adam chronology based on the Carmel *sapiens* finds of fossils around 95000 BC. He is comfortable in his acknowledgement of the limitations of our knowledge in the light of the current scientific (and other) state-of-the-art in this[7,8]. He also thinks that we need to keep open minds on *details* of how man and sin started in the light of more and more mature scientific understanding: *"We treasure tradition not by servile adherence to it, but by, as it were, sitting on the shoulders of fathers and elder brothers who were giants indeed, and thus do we hope to be granted the grace of seeing even further and ever more clearly"*[7]. Many other Christians disagree, and they still believe in a young earth, and a literal six consecutive day creation, and this is a further example of the range there is in Christian beliefs noted in Chapter 7.

SUPERINTELLIGENCE AND WORLD-VIEWS

That brings me nicely to what is for cosmos-views the most pertinent aspect of that key 'interesting and important corner' of cosmos-views I mentioned above – human morality. The Christian doctrine of 'Original Sin' says that all humans are born in a state of sin resulting from the Fall, stemming from a rebellion of our first parents, Adam and his wife Eve, when they contravened a (very small) restriction on the objects they could eat. British Bible scholar, G. Campbell Morgan[9], summarised that probationary aspect of human existence by saying that pristine man (in the shape of Adam and his wife Eve) was *"...put in a probationary position through which he was to pass unharmed to some larger form of existence if his life were to be lived in the union with the God who created him"*. Probation allowed Man the space for the exercise of freewill, but, by asserting this freewill against the probationary limits God had placed on its exercise of it, he rebelled, and the result was the Fall.

Theologians such as Karl Barth, Paul Tillich and others, in trying to get a firm grasp of what this all means, are very interested, in a different way to Krauss, in the role of 'nothingness' – *Das Nichtige* – in the world. Cutting out much of the detail, according to this account, *Das Nichtige*, from which God has separated himself and rejects, takes advantage of what Barth calls 'shadowside', or *Schattenseite* of creation. This includes such 'natural' things as bodily pain or negative correction, say, of a misbehaving child, or having a surgical wound in order to perform some corrective medical procedure, which are 'from the shadowside' of humanity. However, according to Barth it keeps humankind dependent on God for security. Henri Blocher sought to identify the role of evil in Biblical creation in terms of shadowside within the created world. *Schattenseite* corrupted by *Das Nichtige* is 'sinful', and that is how wickedness and other evils are introduced into the world, it is claimed. Factoring in the emotions of shame or exasperation he says sin provokes in us, as we attempt to understand it and as look for its origin and explanation, even believers get into very deep water. Blocher sees the origin of evil as the *misuse of freedom* – first by the personality called the Devil or Satan, and then by Adam and Eve, in accordance with the 'reasonably standard' Fall narrative in Christian theology[10]. As a layman I read these theological conjectures with a lot of interest and respect, some reservation, and a fair bit of puzzlement. When the *sovereignty* of God is brought in we have, of course an 'impenetrable mystery', which is beyond the scope of our present study. However, the concept of sin as being disobedience to a simple command of the designer and creator of the cosmos on the part of our distinguished species has the ring of truth to Christian believers. The historical fact of the Fall, being: *"...the prime evil from which flow all other kinds of evil and misery. It is the voluntary break with God which makes the human race liable to suffer physical evils, and, makes it vulnerable in this world"* [11] makes sense to them. I ignore all sorts of other 'detail' of the 'doctrines' for present purposes, referring the reader to, say, Blocher's or Morgan's work for more, if required.

The main point here is that anyone contemplating the emulation of such processes and phenomena in machines, alongside the corresponding attitudes and values, can see immediately that there are huge difficulties and challenges to be overcome. Moreover, we have adroitly sidestepped the fact that this doctrine of the Fall has an inverse – the biblical doctrine of salvation. Blocher's understanding is that we stand guilty by a mechanism of 'imputation' under God's formal promise to Adam. Mankind's rebellion through Adam taints all of mankind. The primal covenant or sacred agreement Adam and Eve had with God was broken by them. As members of the human race we are said to have a 'legal', covenantal imputation of guilt that is collectively accredited to all of mankind, resulting in individual personal responsibility and guilt. So, under this greatly simplified Standard Model, Adam can be seen as a representative of human covenant breaking. This is a convenient (and considerable) simplification for our purposes here. This is where salvation through faith in Jesus Christ as God's son comes into the picture. It is required if man is to be reconciled to God through an 'imputation' in the opposite direction to that above, and that presents further a big difficulty in sharing our human worldviews with artilects – even in a thought experiment.

Artificial Worship? So, could this *sort of* overall backdrop to human existence be reflected in the cosmos-views of artificial 'agents', such as Watson, Deep Blue, Emily or W*22? And what would be the purpose and advantage of this? I will now present a further, rather unusual, *technology-theologically oriented* view of highly selected aspects of all of this, in order get a slightly clearer handle on the challenge of a few highly selected, abstract concepts, focussing on those demonstrating a *sort of* theological slant (very) vaguely like the 'doctrine' we've just been looking at, and to a much less extent its relationship to that scientific Standard Model. In parallel, we'll bear in mind the practical possibilities for and limits of devices we can plausibly visualise today. If we consider some examples that can be found in movies like *Ex Machina* and *I-Robot* and other media, that go some way towards depicting 'thought experiment' aspects of a 'robot worship world', we will be faced with the *two conspicuous deficiencies* that we've already hinted at. They are far-out *'theologically'* – and this distancing from, say the teachings of Christianity, can, if fact, be an actual requirement if the products are to meet some entertainment level goals – and they have also lost contact with what is *technologically and physically* feasible and plausible.

Consider a well-known robot called Fender Bending Rodriguez (Bender) from a popular television series. He exists in the TV world of *Futurama*[12], set about 2,000 years in the future. Bender is a Bending Unit robot constructed on a robotic assembly line. He is essentially a machine tool for bending metal, but he is also equipped with many interesting 'secondary' functions, either pre-programmed or 'fed' to him in a 'Bending School' section of the robot assembly line, or learned afterwards. We're not given a lot of detail on the technology involved, but it's clearly significantly ahead of our state-of-the-art.

Bender went through different developmental stages as he moved along the assembly from the 'baby stage' upwards, and like a human child he learned to act and speak like an 'adult' robot. Bender has a lot of human-like personality features. Some were set up as part of his design and programmed in, and some were acquired relatively early from his experiences. At times he seems to be caring and is genuinely happy for friends when good things happen to them. He's also rather free and easy morally. For example aspects of his 'romantic relationships' raise questions, and he 'abuses' electricity (in the way drugs are sometimes abused). Important in the context of our consideration of 'original sin' just now is the absence in the Bender world of anything remotely like it or to suggest the existence of a creator of robots, which has godlike characteristics, and perhaps makes behavioural and attitudinal demands of Bender. I haven't seen much of this series, but of interest here is that in one prominent episode[12] in *Futurama*, a 'religious' world-view is dreamed up. 'Robot God' is the ruler of 'Robot Heaven', where, in this simple facetious, unrealistic and somewhat derisive 'religious' picture, good robots go when they die. But how is 'good' defined here? It's left a bit obscure, but it seems to roughly follow some behavioural do's and don'ts from the world of humans.

The alternative, 'Robot Hell', is reserved for robo-baddies. 'Robot God' does not seem to have a prominent creation role with respect to Bender's start-up on the robot assembly line. 'Preacherbot' works or performs in the 'Temple of Robotology'. He does not worship, *per se*, in that episode where Bender has a 'significant' religious experience. At this point I'll give some snippets from the script. The 'conversion' of Bender is, he claims, deliverance of: *"brother Bender from the cold steel grip of the Robot Devil unto the cold steel bosom of our congregation"*. On accepting the principles of Robotology, Bender is 'baptised' in a rather strange way. Later in that episode Bender falls back into his old ways and ends up in 'Robot Hell', is whipped by the 'Robot Devil', who reminds him that he agreed to these 'scary and painful' things when he joined 'that religion' [sic]: *"If you sin you go to Robot Hell for all eternity"*. So, we're not that close to Blocher's theology then!

We can see from other examples of this sort of storytelling that there are visions out there of worlds where there are beings that are super-human, or super-robot, and there are also creators which are mechanical – sometimes human-like superior creatures. There are also some storylines where humans are revered by lesser mechanical 'creatures'. There is an example of such a fictional world which is even further out from current theology and technology than Bender and *Futurama*, in *Star Trek: The Motion Picture*[13], which is a Paramount science fiction film based on the cult TV series, *Star Trek*.

In it, a smart machine, V'Ger, identified by the crew of the *Star Trek* spaceship, USS Enterprise, as originally being a NASA probe, develops into 'a complex system' over a period of time. It heads back in the direction of Earth, destroying things that get in its way and filling up its memory units with huge volumes of knowledge amassed from and about 'races' it has encountered – even more knowledge than Watson has.

It appears that V'Ger has become a 'living machine', which had entered a black hole and 'achieved sentience'. How? That's a good technical question! Anyhow, it moved on with what seems to have been its ('spiritual/theological'?) mission – to learn all it could and bring that information to its creator – ie humankind. It merged those huge quantities of knowledge from the other 'races' in transit to produce 'consciousness', we're told. When some humans on the Enterprise realised what its goal was, they joined with V'Ger, creating a new life form, and the resulting assembly headed off to another 'dimension'. Know what I mean? Some arm waving is needed here, but it is clear that this is all detached somewhat from those two human world-view components discussed briefly above, that is, from science and theology, respectively.

Theological and Technological 'Lacks'. Investigation of the minimal sets of capabilities and constraints that would constitute

SUPERINTELLIGENCE AND WORLD-VIEWS

models of cognition, and patterns of behaviour and causality in an artificial theological 'world' like the robot investigator's in Chapter 5 poses very interesting research questions. Various ways of simulating freewill, for example, could be sought or devised. The imperfections of the fidelity of that system to even simple human concepts in the current ESA system are outstanding however, and there would be many difficulties and details to be explicated if we were to explore or understand both the theological and the technological or physical materialisation – in short, basic difficulties of emulating important aspects of reality. Facts and necessities, and further shortcomings, would become clearer if some experimentation were to proceed beyond our present system. Trying to emulate the real Fall situation might also be in danger of going into *terra prohibita* theologically, of course.

As we noted above, Deutsch for one says that we can't ever get an SI. He extolls the unique capabilities of a pinnacle of creation, humans, although he does not attribute their uniqueness as due to humans being in the image of God, as basic Christian theology asserts. He says[14]: *"There can be no such thing as a superhuman mind...No concepts or arguments that humans are inherently incapable of understanding"*. Deutsch is presumably referring to humankind collectively in this. I noted earlier that I haven't seen anywhere where he explains the evolutionary leap from the smartest non-human animal to the 'standard' human. Maybe the equivalent of a genius-trick, or similar, produced a small advance to get hominids to human level capability. This could be said to be analogous to any leap there would be from human level intelligence upwards. Yet Deutsch somehow rules out the possibility of something similar happening again, at the higher, human-to-SI, level. A standard belief of Christians, for example, is that the first of those two leaps came when God '...breathed...the breath of life; and man became a living soul'.

Theological speculations are interesting, but they need careful handling by the layman. In particular, Tipler was very influenced by theologians whose thoughts raise deep questions. For example, could we: *"universal predictors, knowers, explainers, deciders..."* rely for our persistence on mankind's own provision, and could a move towards SIs lead to worship and service of 'the creature more than the Creator', which was seen as 'a lie' by apostle Paul in a letter he wrote to Roman Christians? The *imago dei* – 'the image of God' – identifies us as a special part of his 'very good' creation. We are a 'little lower than the angels', though, so are the minds of angels above or below SI level? Christian researchers in technology, in particular, would need to think carefully before becoming involved deeply in programmes like those outlined above. Of course, AI techniques are used routinely in applications of computers to improve effectiveness and efficiency. Systems like Watson or AlphaGo, and Airbus, driverless cars and recommender systems on Amazon stand out, but the techniques they use are not in essence very far removed from less automated systems. So concerned researchers should perhaps consider reining-in objectives of achieving SI, and restrict their efforts to producing direct beneficial practical outputs that we're seeing in (eg) medicine, transportation, food-production and other relatively mundane examples of uses of AI.

What is clear is that we're still a huge distance away from having the *Star Trek* or *Futurama*, ie V'Ger or Bender, level of modelling. And we're even further away from the Singularity, and the sort of functionality that Tipler and Deutsch saw as being required for human persistence.

References

1. http://www.geo.arizona.edu/~reiners/geos474-574/Kelley2002.pdf
2. O'Bleness, M., Searles, V., Varki, A., Gagneux, P., Sikela, J.,Evolution of genetic and genomic features unique to the human lineage, *Nat Rev Genet. 13(12)* 2012.
3. Balter, M., Was North Africa the Launchpad for modern human migrations?, *Science,* Vol 331, 2011Download from http://www.springer.com/
4. Armstong, K., *A Brief History of God*, Vintage,1993.
5. Krauss, L., *A universe from nothing*, Simon and Schuster, 2012.
6. Blocher, H., *In the Beginning: The Opening Chapters of Genesis*, InterVarsity Press, 1984.
7. Blocher, H., *Original Sin: Illuminating the Riddle*, Appollos, 1997.
8. Blocher, H., The Theology of the Fall in *Darwin Creation and the Fall*, eds Berry, R. J., and Noble, T. A., Appollos, 2009.
9. Morgan, G. C., *The Crises of the Christ*, Pickering and Inglis, 1945.
10. http://people.bu.edu/WWILDMAN//courWeirdWildWebses/theo1/projects/2001_kazinski/index.htm
11. Blocher, H, *Evil and the Cross*, Appollos, 1994.
12. Futurama, *Hell is other Robots*, episode, May 18, 1999,
13. *Star Trek*: The Motion Picture is a 1979 American science fiction film released by Paramount Pictures.
14. Deutsch, D., *The Beginning of Infinity*, Penguin Books, 2011.

Chapter 9

SI Wisdom and Component World-views.

How do we humans try to proceed *wisely* in any particular matter, and more generally? Well, let's consider what we mean when we say that someone is 'wise'. First of all, it is a compliment. When writing about the loss of openness in contemporary life we looked at in Chapter 6, Allan Bloom[1] suggested that wisdom should be parents' highest aspiration for their children, and most people would agree, I think, that it is something that should be aspired to. Furthermore, it is likely that we would find widespread consent if we surveyed a lot of people on what the main distinguishing features of a wise person are. Most would probably look for significant doses of *'good sense'*, or *'orderliness of thought'* or* 'keenness of insight'*, in someone before giving them this label. Enduring accounts of what wisdom is, and its desirability, eg from ancient literature, usually reflect Bloom's observation. For example in the portrait of wisdom given by 'the wisest man that ever lived', the biblical character King Solomon, wisdom is personified, and is: *"...more precious than rubies, and nothing you desire can compare with her..."*. This passage goes on to say that 'she' has a close relationship with prudence, knowledge and discretion. Ancient sages and kings wrote a lot about it up to and in to the Axial Age from about 800 to 200 BC. However, there is no denying that 'wisdom' is a heavily overloaded term, and it is often attributed rather lightly to individuals.

There is a multi-facetedness of wisdom, which means that there is usually an inherent messiness or muddiness in the making of wise decisions. Among those ancient writers, Socrates tried to spell out what should qualify as wisdom. He seemed to suggest that owning up to ignorance where appropriate was what characterised wisdom. But this cannot be all there is to wisdom. Nor can purely negative reflections about futility. As well as his positive thought on wisdom above, King Solomon in the Biblical book of Ecclesiastes pronounced that *"...all is Vanity"*. These viewpoints are enlightening, though, and, taken along with many others, they are of pertinence to our present focus on world-views and the possibility of machines having world-views. Wisdom affects world-views and *vice versa*.

In human cosmos-views, specialised *component world-views* appear in different domains and they can all affect how wise a person is. Some examples are shown in Fig 9.1. Having looked at the nature of cosmos-views in a little detail and the possible openness flavours they could have, in this chapter and the next I want to look at some component world-views that stand out in a cosmos-view, and I start off with one that illustrates how world-views are important in exercising wisdom in general and not just for the activities of stellar thinkers and creators. Environmental component world-views were backdrops to some of our example cosmos-views in Chapter 4, and we've met the Transcendental world-view that was seen to be an important component when we were discussing the future of the cosmos. I want to pick up first of all on that very brief consideration of environmental outlooks, and accompany it in with a quick look at wisdom. In the next chapter I'll look at three other conspicuous components – Science and Philosophy alongside Technology – which have a significant impact on a human agent's formal understanding of nature and matter, and things in general. The first two of these are clearly linked to the Transcendental world-view discussed in Chapter 7, and they are in fact sometimes seen as being in opposition to it. Technology is a key component world-view in our context, and it is related to Economic, Social and Political world-views, which I will also briefly mention.

The environment is, most would agree, important in its own right as well as in its relationship to other 'component' world-views. It is concerned ultimately with the habitability or otherwise of our Earthly 'spaceship'. So, by way of introduction, I want to use attitudes to the environment to illustrate how I believe wisdom and world-views are intertwined. My suggestion is that SIs would need to have both wisdom and world-views, so an obvious question is: In what way are they linked? To tackle this question, consider a debate that's widespread these days on the process of 'hydraulic fracking'. This is a process that is invoked by drilling an access pipeline deep

SUPERINTELLIGENCE AND WORLD-VIEWS

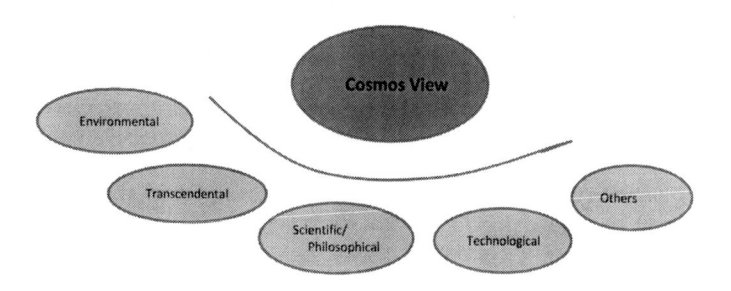

Fig 9.1 Some Cosmos-view Components.

into the ground, through which fluid is pressure-injected in order to fracture shale and release gas, which is valuable as a fuel. This process has a lot of promise. It's said that it has the potential benefit of producing large quantities of natural gas. There are risks, though, to humans and to the environment. A large number of chemicals are used in the pressurised fluid, some of which are agents directly involved in causing cancer and some are toxic. Contamination of ground water, and then maybe drinking water, can potentially cause neuro-biological, sensory and respiratory problems. So the question is: is shale prospecting and gas extraction worth it? Your answer to this question will depend heavily on your environmental world-view. Disagreements on how serious the environmental risks are, on what the associated problems are, and on how we should tackle these, often cause quite deep divisions between people who have competing environmental world-views, as we'll see in a minute.

Good Sense and the Environment. It is worth looking at that notion of 'good sense' a little more closely, especially if we aspire to having machines with it. A prominent feature of good sense is that it involves some notion of *'completeness of thinking'*; ie making sure that we consider all the pertinent factors, or as many as we can know, about the matter in hand. Intimations of *'measuredness'*, and *'balance'*, are also carried by 'good sense', and they are related to carefulness and restraint in considering issues. All respondents to our imagined survey on wisdom would say, presumably, that good sense, perhaps called something else, should be possessed by a person at an exceptional and sustained level before we'd call them 'wise', and that it's desirable to employ it to some considerable degree in all sorts of decision-making such as in choosing what to do next.

So 'being wise in a particular situation' is taken here to mean using good sense in order to be successful at picking the right thing to do or think, and being disposed to commit to that selection by doing or believing accordingly and appropriately (without raising one's voice). This means that someone who tends to be recognised as being good at choosing sensible goals in life in general, or in particular projects, and then choosing the best ways of achieving them, should be called 'wise'. And it is my belief, for reasons that I will reveal, that wisdom is closely related to one's cosmos-view, and, to varying degrees, to one's more specialised component world-views. An implication of this would be that realistic, or credible, world-views are essential if an agent is to display wisdom – ie to be wise.

Using this informal take on the meaning of the word 'wise' in the meantime, it is easy to find contrasting judgements on what is considered wise where different world-views are considered. In cases like the 'to frack or not to frack?' debate, we have to choose between the predicted bad effect of some action on the environment and some basic potential benefit, or 'good', for humans that the action would bring about. Your answer to our question on how to proceed wisely on the fracking issue would depend heavily on the details of your environmental world-view. Whether it is in fact wise or not, according to our working definition – choosing well and doing accordingly – would then depend on how realistic that world-view is.

Your particular environmental world-view might, for example, be close to those 'good-life' outlooks in some of the illustrations we thought about in Chapter 4. You might then emphasise, for example, the fact that living things depend on nature in a very important way, so consideration of even the positive consequences of measures like fracking that aim to help with the economy, worthy as they sometimes are in their own right, are *secondary*. They should be subjected to very careful scrutiny to see how nature-friendly they are. Under this view, individuals and communities must always integrate consideration of nature carefully into all the ways in which we think and live. Now one thing this requires for sure, is *knowledge*. We need to understand, in some detail, the human condition and the ways in which we think and live and depend on nature's bounty. This in turn, includes an understanding of how natural systems maintain themselves and propagate. Operations involved in fracking, for example, are known to threaten to degrade one's surroundings, near and far, and, given the 'good life' stance we're adopting for argument at this point, that sort of operation should be discouraged. Having this view would also dictate that further knowledge of and about resources be sought, as stewardship of those resources is important. The focus under this particular environmental world-view is on the fact that if the resources we can access are in fact *limited* – a view opposed by people like Deutsch, of course – then humans have an ethical responsibility not to squander them, and to use best practice in their management. Using good sense *here* would entail adherence to this view and use of our knowledge conscientiously and measuredly in making decisions, and when acting accordingly.

The cosmos-view we looked at in Chapter 7 when considering Deutsch's ideas, incorporates a different working belief – that economic and other growth can be effectively *unlimited* in the accessible world, and that over-concern about natural resources just holds us back. Under this model, human advancement and self-expression benefit through putting 'nature's bounty' to good use for these, and especially for making machines with impressive functionality, are not secondary, but paramount. It is sometimes claimed by people taking this stance that mankind should use genius to 'do better than nature', and that we should recognise what is the reverse of the previous precedence order and put humans' interests ahead of those of the rest of nature. We should, of course, still be measured in our dealings with nature as a *secondary* task, but our main consideration under this view is to know how to meet our increasing needs and wants, recognising the 'intellectual heroism' and smart thinking required to improve both knowledge and technology. Leo Marx[2] quotes 'a visitor to Yale in 1831' who extols inventors of his era in a question: *"...is it quite certain, that anything short of the highest intellectual vigor, – the brightest genius, – is sufficient to invent one of these extraordinary machines?"*.

Which of those two versions of environmental world-views should an SI be given? The designer's own world-view will loom large in this choice, but there are also clear SI-specific aspects of this, such as: are those human's interests important to the machine? The choice of world-views will affect the wisdom of the SI. An SI might pose some unexpected queries – eg is the prudence of preservation and conservation of the environment irrefutable?

Wisdom. Measuredness, completeness and acting appropriately in the 'fracking case' would be in subjection to an agent's environmental component view. So what would be *considered wise* would be quite different under each of the two versions. To discover what *is in fact wise*, we would then have to know which world-view is more 'correct'. Remember I have been assuming that there is a 'real world' in which there are absolutes that we can relate to. The wisdom of a human agent is thus heavily dependent on the *quality* of the agent's world-views. How we value some particular things, the way we feel about them, and the ways that we tend to act, are all interrelated with our values and life-goals, and to the measuredness and completeness of our judgements and insights, for example. So quality of wisdom is affected by any ways in which these goals are personal and subjective only, or otherwise relative, or even wrong. Furthermore, the exercise of wisdom involves some reflection on the vagaries and uncertainties of life. A healthy attitude to risk is required, implying strong will power and a subduing of doubts in favour of higher 'rewards'. The influential Brazilian education writer, Paulo Freire[3], wrote that: *"...our condition as extant beings subjects us to risks";* and we must: *"recognise the risk...in order to proceed effectively".* A wise person will acknowledge the inherent uncertainty of life, and be able to reason and act effectively in spite of this, possibly using an approach where disagreements are resolved through measured discussion, while maintaining emotional stability. We would of course like all of our working knowledge to be absolutely certain, and to correspond exactly to the real states of affairs. We'd like all of our cognitive, reflective, and affective or compassionate deliberations to always be impeccable. But can this ever be claimed? We humans have to deal with this uncertainty all the time, and so would SIs, as we'll see in Chapter 12. Handling uncertainty at a naïve level should be easy enough for future robots, but the task of fabrication within robots, for example, of emotional and value-related aspects like those present in human decision-making, is daunting. So prospects for 'wise' robots are quite gloomy. The equivalents of the capacities that humans have allowing us to commit to actions in the messy and muddy sorts of situation we're faced with seem to be essential for the machines we're talking about. So they would require that desirable characteristic of wisdom.

Whatever our preferences and precedences happen to be, there is, by common consent, an *active* side to wisdom. There is usually something *to be done,* or at least, *to be committed to,* when we are prompted to use it. I'll assume that wisdom involves putting one's knowledge about how to live well into action. Then there will be no 'wise' individual who knows all about living well but doesn't live well in practice[4]. Good judgement, incidentally, is judgement that is "reliable, sound, reasonable" – like our 'good sense' above. Mental deliberation on the truths and falsities of situations encountered in life often involves consideration of practical matters and physical bits and pieces, and the exercise of wisdom is very often taken to

SUPERINTELLIGENCE AND WORLD-VIEWS

have practical outcomes – like making a choice between some tangible things, FAPP. Everyday things that are important in the individual's approach to life and work determine his or her priorities. They provide a basis for the measures needed to evaluate success and identify failure with respect to how things are turning out. So we see again the importance of world-views – in particular, *evaluation* – in many decision scenarios, and that it is a key issue for understanding, explaining or ordering things in the world. As noted earlier, we humans usually appeal, however automatically or unwittingly, when 'sizing up' or ranking alternatives, to a set of values that we have acquired from school, home, religious tradition or community, or other conditioning influence. Incidentally, in the opposite direction, wisdom can direct *the selection of values*, adding one more sort of mutual recursion to the list given earlier. As life goes on we might use wisdom from time to time as we adjust values a little, and, vice versa, the values affect our wisdom. I ignore that complication here, and for clarity at this point think of our values being pretty well fixed. We'll look in more detail at values in Chapter 11.

Values and goals, and therefore, by association, wisdom, are closely related to *emotions*. When outcomes match values nicely, an appropriate emotion such as contentment is triggered, but when they don't, sadness or anger emotions may be triggered. When making nearly all decisions, the set of values that the deciding agent holds work together with its emotions to facilitate choice or to finally commit to a decision or belief. Deciding is rarely based on Spock-like, purely intellectual activities with ultimate objectives being to discover the truth, or to make an optimal move, *in vacuo*. Rather, gut feelings, simple prejudices, curiosity, anger, joy or sadness, and more complex emotions, often come into the picture[5] and I'll return to that in Chapter 13. For example, curiosity, which can be seen as a value or an emotion, is something we are very interested in the AI world, especially in exploration scenarios. It can be defined as an emotion that directs attention strongly toward some incoming perceptions. Especially interesting is where we investigate something novel, perhaps a sudden change in the environment, or where we seek the cause of something that actually happens being extremely at odds with what is predicted to occur. As the agent responds to curiosity triggers, new perceptions can be accumulated. Taking the advice of experts 'hook line and sinker', with passive assent, is not to be recommended in many situations. Better to collect and 'weigh' the pros and cons, costs and benefits, of each. Then, when all the relevant information we can get is in, or we have run out of time, or maybe (usually?)

earlier if our emotions are of a certain kind, take the plunge, often depending more heavily on the 'feel' of the situation than we like sometimes to admit[6]. For emotions, *meta-goals* affecting our world-views such as 'to keep a positive emotional state' can also be set.

Wisdom Paradigms. Because of the interdependence of wisdom and world-views, it is important in our study of world-views to look a little more carefully at what various people have said about what wisdom is. Despite what was said above about a survey producing broad agreement on some of its features, not everyone has a completely clear view on this. At a meeting I attended recently, an academic colleague from another discipline, having a PhD in a science subject from several decades ago, responded, admittedly after a rather short time for reflection, to a request for a definition of 'wisdom' by saying that for him it is synonymous with 'having knowledge'. But by our discussion above, we see this is not enough. Knowledge, valid reasoning and truth are of course conspicuously associated with wisdom, but other components of the concept may stand out less immediately, and yet they may be more prominent when we do have time to reflect on this. However, even among those who have sought to spell out the character of wisdom, not all have arrived at the same conclusions – but there is a substantial overlap between their respective findings.

You might be surprised to learn that there is such a thing as 'The Berlin Wisdom Paradigm'[7]. My friend's quick response is backed up to some extent in this Paradigm, due to the fact that knowledge, such as: *"rich factual...and rich procedural knowledge about life"*, alongside more contextual knowledge, is a conspicuous requirement. Indeed, wisdom is said to also include: *"expert knowledge in the fundamental pragmatics of life that permits exceptional insight, judgment, and advice about complex and uncertain matters"*. However, alongside knowledge, the Paradigm includes many other things as requirements for wisdom, such as 'expertise in the conduct and meaning of life'. This latter rather opaque phrase serves here to indicate the strong link with cosmos-views and world-views; criteria suggested include having a wide-angled view of the 'full context of life'. There have been a number of other attempts to get a 'wisdom paradigm', a way of looking at wisdom like the Berlin one, and these do have a large degree of agreement, so I will only give a sample here.

The findings of another research group which has been looking at these issues for a long time gives three related key components of wisdom, bringing emotion clearly into the picture: *cognitive* (to do with desiring to know the truth, taking into consideration uncertainty, yet remaining able to make decisions in spite of this), *reflective* (concerning looking inward at one's self and being open-minded with an ability and willingness to look at things from multiple perspectives), and *affective* or *compassionate* (remaining positive emotionally in the face of adversity and taking into account the well-being of others)[8]. This group at one stage replaced this trio of components by six more detailed components which we list for completeness here: pro-social attitudes and behaviours; social decision making and a pragmatic knowledge of life; emotional stability and relatively constant condition; reflection and self-understanding; being able to live with others who have different values; and acknowledgment of and dealing effectively with uncertainty and ambiguity[9].

In yet another study[10], most of the experts who were polled agreed that, among other things, wisdom is: *"uniquely human"*. Presumably they weren't considering the possibility of any future SIs being wise. They define it as: *"...a form of advanced cognitive and emotional development that is experience driven; and a personal quality, albeit a rare one, which can be learned, increases with age, can be measured, and is not likely to be enhanced by taking medication..."*. Its link with active experience is also captured in an encyclopaedia article as we saw above : *"S is wise [if and only if] (i) S knows how to live well, and (ii) S is successful at living well"*. For such reasons many regard it, almost by definition, as something that comes only with maturity. Other researchers have looked at how personal human attributes such as age, education, race and gender, impact on wisdom[10]. The ability to live with contradictions and alternative hypotheses or conjectures is often thought of as being a negative, but this ability can be a positive, for mature people as well as for youngsters, as we'll see in Chapter 12. Notice that most of these suggested components of wisdom can be seen in our earlier examples of human world-views, like Cobbett's or Livingstone's.

For completeness, I should mention a different approach to the understanding of wisdom of those who see some promise in the idea of aspects of wisdom as being 'uniquely human' in the sense of its being neuro-biologically determined, and mapping nicely onto neural substrates[12]. For example, a human's *medial prefrontal cortex* is implicated in determining how intrinsically attractive (eg eliciting joy) or off-putting (eg eliciting fear) particular circumstances, happenings, or objects are to the person. Pro-social attitudes are linked with the 'reward circuitry' in the *ventral striatum* and *nucleus accumbens*, areas associated with addiction. Regulation of other specific brain regions is involved in decision-making and value relativism. I will be largely ignoring this for SIs since our 'SIs' won't have naturally built-in *nuclei accumbens*, or any other of those body parts just mentioned. We are not considering the Whole Brain Emulation approach to try to materialise our SIs. That's a direction I personally do not see as being very promising, and it's outside our scope.

Whatever the final definition one chooses for *wisdom*, it has to be emphasised that what would be *considered wise* in real situations would again be quite different under different choices within each of one's relevant component world-views, some of which I look at in Chapter 10. If there is an absolute reality, as I'm assuming, the implication of this is that to discover what *is in fact wise*, we would have to know which world-views and cosmos-views are *correct*, and if give them to any future SIs – if we could!

References

1. Bloom, A., *The Closing of the American Mind*, Simon & Schuster, p58, 1987.
2. Marx, L., *The Machine and the Garden*, Oxford University Press, 1964.
3. Freire, P., *Pedagogy of the Oppressed*. New York: Continuum Books, 1993.
4. Ryan, S., "Wisdom", The Stanford Encyclopedia of Philosophy (Winter 2014 Edition), Edward N. Zalta (ed.), http://plato.stanford.edu/archives/win2014/entries/wisdom/
5. Baltes, P. B., & Staudinger, U. M. (2000). Wisdom: A metaheuristic (pragmatic) to orchestrate mind and virtue toward excellence. American Psychologist, 55 (1), 122-136. Accessed from https://www.mpib-berlin.mpg.de/volltexte/institut/dok/full/Baltes/wisdomam/index.htm
6. Tversky, A., Kahneman, D., Judgment under Uncertainty: Heuristics and Biases, *Science*, New Series, Vol. 185, No. 4157, 1974.
7. Banicki, K., The Berlin Wisdom Paradigm: A Conceptual Analysis of a Psychological Approach to Wisdom, *History & Philosophy of Psychology*, 11(2), 2009.
8. Ardelt, M., Wisdom as expert knowledge system: A critical review of a contemporary operationalization of an ancient concept, *Human Development*, 47, 2004.
9. Bergsma, A., Ardelt, M., Self-Reported Wisdom and Happiness, *J Happiness Stud*, 13:2012.
10. Jeste, D. V., Ardelt, M., Blazer, D., Kraemer, C., Vaillant, G., Meeks, T. W., Expert Consensus on Characteristics of Wisdom: A Delphi Method Study, *The Gerontologist*, Oxford University Press. Vol. 50, No. 5, 2010.
11. Meeks, T. W., Jeste, D. V., Neurobiology of wisdom: An over-view. *Archives of General Psychiatry*, 66, 2009.

Chapter 10

Science, Philosophy, Advancing Technology and Progress.

Giving some sort of very limited component world-views of the cosmos at environmental level to the ESA robot investigator in Chapter 5 might just be possible. However it is clear that reaching human-level agency using such apparatus by bootstrapping alone would be well-nigh impossible. The same can be said for other components of a cosmos-view. A scientific world-view would be one of the component world-views that would be very useful for a robot explorer, as we'll see just below. In humans it relates closely to one of our 'big two narratives', and some would say it is the predominant driver of cosmos-views these days. Philosophy is also important for humans, as it is so close to cosmos-views that some people can't distinguish between them. Technology, the third component we look at in this chapter, is central to our whole subject in this book.

Science and Philosophy. A variation and simplification of the ESA robot system can be used for illustration as we direct our attention to some more specialist ingredients of cosmos-views. Suppose a small mobile robot is introduced into a very limited world, which takes the form of an essentially empty arena – say an eight feet square room. It can move around and it encounters new objects (say lights and doors) in its world, which 'behave' in some ways – like the prey in the Chapter 5 robot experiment. It is interesting to think of some built-ins the robot might be given. The robot could be made to *sense* features of the objects it encounters and to *react* in a pre-programmed ways to its perceptions.

Suppose our robot is also programmed to explore, and that it is given certain 'appetites' and 'aspirations', and basic learning and reasoning ability. We are assuming that it won't be given *all* of its knowledge *ab initio*, play-back, so a large chunk of detail in its cosmos-view can be obtained and changed, taking a 'bottom-up' approach – knowledge acquisition 'from experience...through exploration of its world'. But although the explorer will not have 'on board' built-in, full and fixed environmental, transcendental or other specific component world-views, such as the scientific, technological or philosophical world-views, it will have those pre-programmed basic building blocks. The designer might also select additional basic features from the literature and expert opinion in the form of existing concepts, models and functionality for inclusion in the robot's repertoire, and supplement them from experience. In limited implementations based on this scenario, the robot's approach to seeking discoveries in its world has, by default, been *sort of* scientific. This in turn invokes adoption of certain *sort of* philosophical standpoints with respect to issues of epistemology and metaphysics that arise from such exploration as it is carried out. A very basic scientific approach gives a good way of starting to look at the world of time, space and sense – important for robot explorers.

The limitation of any picture of scientific discovery in this scenario is acknowledged at the outset. Scientific knowledge includes much insight about the world that is not directly accessible using our given *human* senses, never mind a robot's. The Higgs boson was hard to detect until recently. Moreover there are aspects of scientists' activities and thought processes that are beyond the simplistic empirical-theoretical model normally associated with the scientific method. For example, a *faith-stance* of sorts seems to be widely taken in human scientific endeavours. The moral philosopher Mary Midgley, whose 1985 book *Evolution as a Religion* stimulated my own thinking on this many years ago, and who opposes attempts to present science as an ideal answer to all questions we might ask, says[1], linking up with the thoughts of Tipler and Teilhard: "*...progress (smoothly dovetailed with evolution) has increasingly appeared as an escalator, powered by our own remarkable abilities and bearing us – perhaps with cybernetic additions and perhaps becoming immortal – reliably on towards a distant and mysterious Omega Point*". She says[2] that faith is also often invoked in various 'seed-beds' of religion which have 'characteristically religious' elements like: "*...priesthoods, prophecies, devotion, bigotry, fanaticism*" and even notions of: "*heaven in the future*". Furthermore this faith manifests itself in a current Zeitgeist that is not always thought through fully. There are some aspects of historical research that are convincingly 'scientific' to the uninitiated, and are wrongly taken as being exceedingly reliable by those used to encountering daily evidence of the reliability of the work of scientists, in the operations of laptops, cars, and aeroplanes. But there is a subtle difference between the *experimental, nomothetic* (from Greek '*nomos*' meaning 'law') or *operational science* underpinning these devices, using a very strict empirical-theoretical framework, like the 'understanding' that our robot achieves, and such '*historical science*'. The palaeontologist, Henry Gee, a senior editor of the journal, *Nature*, has picked out flaws in a domain of historical research with which he is familiar. He claims that the details of the sequencing of man's development over the ages is subjective, and: *"They can never be tested by experiment, and so they are unscientific. They rely for their currency not on scientific tests, but on assertions and the authority of their presentation"*[3]. Trying to determine how past evidence, signs, or other data can be pieced together in order to explain events in the past in fields such as in Evolutionary Studies, Forensic Science or Archaeology is not nearly as reliable as standard operational methods in research. So scientific pursuits are not as simple as is sometimes made out, and our mechanical robot 'scientist' has a lot of shortcomings.

The prodigious feats of mankind in transcending the time limits imposed by the brevity of one human life, and combining the memory capacity of very many human brains, do have to be acknowledged. Collectively as a race we understand our immediate environment in remarkable detail and can produce fascinating models of the remotest locations in the cosmos using astounding insights that individual brains have uncovered. But this does not warrant the 'characteristically religious' stance taken by those with a 'scientific *cosmos*-view' or *scientism* of the kind Mary Midgley objected to. Our robot's exploratory *modus operandi* as a naïve scientist should not to be confused with the robot's having scientism as a cosmos-view. Proceeding in that particular way does, however, indicate what could be seen as a *component* of a cosmos-view. In humans this component specifically supports some important aspects of the more objective side of the representational aspects of our human cosmos-views. A science-based approach to knowledge acquisition, many would say, strongly impacts current world-views, and scientism *is* sometimes taken as a significant default for a human cosmos-view, with its reliance on just one of our 'big two narratives', in its own right. However, many, and hopefully most, thoughtful people, including numerous scientists, as we saw in Chapter 7, have a broader cosmos-view. They believe that while the use of the scientific method is a positive mode of reaching knowledge, it is *only one* of the ways of doing this – contributing as it does to *only some* of its aspects. There's a stark contrast between the 'scientific' *modus operandi*, which our robot explorer *sort of* parallels, and scientism which adheres strictly, exclusively and 'religiously', to the empirically verifiable, and plays down the importance of, other philosophical, transcendental, metaphysical and epistemological knowledge-acquisition methods.

If we want to engage in completely open-ended, unlimited inquiry, the very method of science itself must finally be seen within a wider context, and this induces us, naturally enough, to focus for a moment on another component of cosmos-views we're looking at just here, the Philosophy component. In particular I want to consider the topic of *epistemology* – the study of knowledge – which clearly has to do with explaining and understanding. I'll be saying more about knowledge in Chapter 12, but it seems to me to be clear right away that the discovery of knowledge should not be limited to any particular methods of science, as abstract and subtle *metaphysical* questions arise that are beyond the reach of scientific methodology. For example, the issue of how the brain gives rise to experience *at all* appears, at present at least, to be a (big) metaphysical question. Max Planck a key pioneer of quantum physics said that, because, *"in the last analysis, we ourselves are part of nature and therefore part of the mystery that we are trying to solve"*, science cannot tell us everything about consciousness, for example. And of course, 'ultimate' questions about things like various origins, such as of life and of the universe itself, or even the existence of God are reckoned to be beyond the scope of sense data.

Consideration of cosmos-views, that 'top-level view of the collection of things in their entirety', brings under a single blanket all the issues that are raised by epistemological and metaphysical reflection on reality. Questions arise straight away about assumptions and presuppositions upon which this epistemology is based, and this is what metaphysics deals with. Are they arbitrary – unargued, and perhaps based only on opinion – or are they based on some absolutes and established by some rational processes? What are their preconditions? How consistent are they? I don't want to emphasise those sorts of questions here, though, as I simply want to focus upon the scenario used in the experiment described involving our autonomous, cognitive, mobile robot explorer. The way it explores its 'world' is an illustration of the scope of a quasi-scientific approach to describing and perhaps explaining physical and other 'real' entities of interest. The robot is not, however, concerned with 'what is

SUPERINTELLIGENCE AND WORLD-VIEWS

real?' It could be said to work according to a naïve *realism*, rooted in an understanding, like the 'man in the street' one – that senses provide us with direct awareness of the world. Entities are taken to have attributes – properties, such as colour and size, name and date of birth – that are as they are perceived. Remember, we're assuming that there is an absolute reality. That reality goes beyond the arena in which the robot works, and includes the human researcher, for example (incidentally leaving room for 'miraculous interventions' by the researcher!). In the meantime we simply let our envisaged mechanical explorer proceed as a naïve scientist, focussing on the aim of allowing it to get knowledge of things that it can sense physically. Exploration by human agents has succeeded in doing this to a very large extent, and hopefully robots using the same approach can do this pretty well.

Now, another *sort of* philosophical stance could also be said to be seen in this approach, called *instrumentalism*. Scientific theories, including naïve ones, are, it says, simply useful but disposable instruments for helping us to make sense of and to predict results of observation and experiment. The objective of a useful robot won't, on this view, be to read off features of reality, but it will be equipped to get information that helps it to adapt to, control or transform its sensed world. It is interesting that, let loose in its very simple environment, just one of those two philosophical stances – naïve realism or instrumentalism, would be given to it, perhaps tacitly – without being made directly 'aware' of it – by the designer or programmer.

As I hinted above, many human experiences are beyond the scope of this simple 'physical exploration' approach. 'Objective knowledge' obtained using an empirical approach does not offer a lot of help when consciously examining our own thoughts, but it's also deficient when matters such as specific, identifiable values, things like Plato's forms and all things supernatural are considered. The question arises, of course: how can an explorer know whether statements about those things that can't be tested through sense experience are true or false? Taking the instrumentalism stance for SI, this would not be a big problem.

What more can be said about our smart robot's *general outlook*? OK, it is a naïve scientist, and this means that there must be 'scientific' aspects to its perceptions. Any other features of aspects of its cosmos-view could be explicitly given in a basic form and perhaps developed somewhat in the form of a short series of paradigms. The concept of cosmos-view is often taken to be more or less identical with that of philosophy. However it is taken here to be substantially broader in meaning than the concept of formal philosophy. We often talk about the 'philosophy of life' of a human, and this might get us a little bit closer in meaning to it – a less formal attitude to issues like our mortality, obligations that we have to our 'world' and what we, as humans, should be doing with our lives. A human cosmos-view also concerns 'timeless' characteristics of humanity, such as, perhaps, intrinsic values, and we'll look at those in Chapter 11, and there is an apparently intrinsic need to satisfy enquiring minds and appetites. It also has the flavours we've met before to greater or lesser extents – openness, or not, purely material or not, and perhaps transformative or not. Any non-human agents, including our robots, would have to have equivalents, where appropriate, if they were to have cosmos-views. Human philosophy *per se*, in contrast, is understood here as a label for abstract, critical inquiry at a very general level, and methodical speculations addressing the highest questions about the reality of the world that need long, principled deliberation. Cosmos-views are richer and broader than this, but in the absence of a special Philosophy component they are not so deep or formal. They might come in part from reflection on the sort of understanding acquired via that more formal philosophy, but they would combine it with a much more extensive sense of commitment and value, taste and judgement, and with strongly-held *principles* that govern actions FAPP. Both philosophy and cosmos-views are, of course, necessarily informed by reflecting on accumulated experiences of various kinds, and also by culture and inherited innate tendencies. A Philosophy world-view, as a component of a cosmos-view, is rooted in rationality and is thus also aimed at a kind of universal validity, and all of the well known, hard, general philosophical questions are of paramount importance, and they are persistent. My own position on this is close to what Vidal and others refer to as the: *"world-view crowns philosophy"* stance. By this they mean that a cosmos-view (in our terminology) is 'the highest manifestation of philosophy'. Vidal's 'definition' of a cosmos-view[4] improves on Freud's. It's a coherent collection of concepts allowing us: *"to construct a global image of the world, and in this way to understand as many elements of our experience as possible."*. A cosmos-view without a Philosophy component is considered to be focussed too much on practicality for formal philosophers. We can look at an instance of that component, a world-view within the global cosmos-view, as belonging to 'the class of philosophical world-views'. This Philosophy world-view component can then be as formal as desired, being grounded in rigorous rationality.

SUPERINTELLIGENCE AND WORLD-VIEWS

It is interesting to note that, within *any* single world-view as a component of a given cosmos-view, several different philosophies could be accommodated for different people. In a single community there could be farmers with a particular world-view who have a predominantly *pragmatic* fix on their work and life in general, scientific researchers with that same world-view who adhere to an effectively *positivist* view-point in theirs, and even *existentialists* with the same world-view. Also consider the example of the 'pre-Bloom world' of earlier chapters. Many people having the original 'open-minded' cosmos-view who adhered to the same philosophy may have displayed other forms of diversity, for example, a range of political allegiances. Close questioning would probably have revealed that there were both socialists and conservatives in that world, and their respective detailed philosophies would reflect this.

A cosmos-view does not change the ultimate reality, but clearly it should correspond as closely as possible to the absolute reality we're assuming. This would also be a requirement for any future 'true SIs'. And there are other similarities with the human equivalents. For example, our cosmos-views should be 'up-front', and this should perhaps be required for SIs. If people are not aware what my cosmos-view or a particular world-view is, they will likely have to make a lot of guesses about what I mean when communicating with them, and they might misunderstand why I act as I do in certain circumstances. Individual world-view holders are (of course) limited by their inherent capabilities and lacks of capabilities, and by the extent of their exposure to guidance and good experience. Corruption can also raise its ugly head, so that some world-views may become expediency-based. This is considered widely as being undesirable.

Turning again to our inquisitive robot, let's suppose its exploration proceeds incrementally within its heavily-constrained robo-world. It's clear that the 'physical exploration prototype' must be extended greatly if it is to get anywhere near having such the human characteristics as wisdom, deep philosophy or even, as Ava presumably hoped, having some pragmatic capability of survival in the physical universe. For example, Vidal and others[4] in the A-schema we've considered in Chapter 3, say that a cosmos-view should comprise: a descriptive *model* of the reality of the world; an *explanation* of the world; a *prediction* of what the future holds; a stipulation of what we can and should *do ethically*; a theory *of action* – on how to attain our goal; and finally, an *epistemology* – what is true and what is false. The flavours associated with a particular cosmos-view can then be used to refine it further, and even to partially answer questions about the structure, dynamics and origins of the cosmos-view itself.

Human wisdom is built on such foundations. So, if we are ever to give the robot explorer, or an SI, a cosmos-view, and with it the potential to be wise, we are faced with very, very hard problems.

Technology and Progress. Creativity, openness, values and purpose should all accompany developments in technology and science. They were particularly important for many of the advances in recent decades and centuries, and collective effort characterises the approach taken to this 'progress'. We can learn much from all the changing social, environmental, economic and political factors in the experiences of our forebears, but of most interest in our present context is that technology changed over the epochs. Initially things moved forward as the materials that tools were made of changed. Stone and wooden tools were supplemented and largely superseded by metal knives, axes, and parts of ploughs and weapons. Pottery storage vessels also enhanced the efficiency of important kinds of production.

Settled groups started to appear and this allowed larger herds of domesticated livestock to be raised, crops could be grown more intensively, and specialisation increased. Technology doesn't explain everything in the evolution of populations, but for a long time subsistence technology had set limits on what was possible in a society, and it seems that advances away from the most basic technology had to predate other strides forward. Tools to help get rid of the heavy and mundane would have had a great attraction, especially to all smart, lazy people. Moreover, technologically advanced societies are more likely to transmit their social and cultural characteristics to future generations.

Humans are not simply singleton explorers with a built-in inquisitiveness about their worlds. We are social animals who cooperate in order to maximize 'success'. Success can, of course, be defined or assessed in many different ways, from simple survival to full exploitation of opportunities that arise to further various goals. The ability to mentally wear someone else's shoes – to see things from the other person's stand-point – enables us to act for the common good, for surviving, thriving and achieving as a species, using our sophisticated critical thinking skills. In the *Ex Machina* movie the lack of empathy Ava seems to display with respect to the humans she is involved with might not be critical in the short term as she sets off to explore the world and

SUPERINTELLIGENCE AND WORLD-VIEWS

observe busy road junctions without back-up from other agents. However, although she will probably be able to look after herself for a while, this will not last forever.

Human societies can seem to be a setting for co-operation, but they can also harbour perpetual competition if resources for the conducting of well lived lives and custody of the environment come under pressure and demand exceeds supply. We therefore have to have the cognitive and other capabilities that facilitate the acceptance of systems that contribute to the management of our affairs at a collective level. The management of the means of survival is, of course, a basic platform from which the struggle for other kinds of success is carried forward, and that evidently requires an agreed, collective character if it is to work. So, for example, the openness, values and purposes of the society to which we belong will probably set standards that should be recognised and adhered to by its members. Society, then, affects our world-views very powerfully. Incidentally it seems to be unavoidable that, in turn, any SI world-views, and even their design and development if it happens, in the early days at least, would be subject to, and heavily influenced by human society and any other aggregates of agents.

Technological factors and innovations also remain prominently in the picture and have a huge impact on the way the people live, work and learn, and they are closely intertwined with the scientific narrative and scientific genius. They affect communications, travel and energy-related matters. There are other factors impacting life and 'progress'. The distribution of peoples' resources is often carried out within a system of economic and political rules. For example, these days government infrastructure and government taxation policy play parts in this. Environmental impacts also have to be taken into account involving as they do ecosystem factors like air, water, soil and food. In the history of mankind here is a pattern of constant change in society, and within a society, social factors such as obligations of various kinds to other people play their parts. The compound of consumption, demographics, lifestyles, collective values and individual values, is therefore very complex, as it has to deal with great variety among individuals who each have their own personal emotions and appetites. Those five characteristics or 'dimensions': social, technological, economic, environmental and political, taken together, provide a useful framework sometimes called the STEEP framework to accompany the scientific and transcendental narratives for checking developments and interactions in various scenarios. Each of those factors could be seen as an added component world-view, but we place emphasis on technology, the 'T' in our acronym, as this is the factor I am most familiar with and the one that is of most relevance here. The attitude of individuals and communities to technology is a conspicuously important feature of human life. The stance taken on technology is especially interesting in the context of cosmos-views that contemplate SIs, impacting on the basic character and feasibility of related techniques.

We are interested in this book in, among others things, the benefits offered by the development of SIs, and in whether or not we will really need their additional functionality and the envisaged increase in freedom in the future. The potential of the practical fall-out for safety and prosperity is clear from some current applications of AI. Such 'progress' can be seen as another sub-narrative – that of well-established and continuous technological advance – especially since Cobbett's time, and particularly in the last two hundred years. The leading edge of the associated 'degrees of freedom increase' has been progressively advanced by overcoming natural constraints on human action by the use of increasingly sophisticated tools. The corresponding unfolding story line could almost be taken as another narrative on the level of our 'big-two'. For humans and SIs like W*25, a technological world-view would, of course, have to cover attitudes towards any next generation of SIs.

Positives and The Down Side. It was booming industrialisation that really set the ball rolling towards any SI development. The steam train, the Spinning Jenny, and the Jacquard loom, and other powerful machines paved the way in the 18–19th centuries, and they freed people up, in a relative sense, notwithstanding some negative thoughts and comments from Cobbett, for example, on the ubiquitous paper mills and other factories that he saw appearing in the middle of his life, when the industrial revolution had hit country dwellers hard. The push for progress was helped by a modest adjustment downwards of the level of social inequality among the members of society as material rewards were spread wider. Moreover, premature deaths and afflictions such as hunger were drastically reduced in many societies as a result of industrialization. As agriculture became more automated and intensive, the quality and quantity of food improved, and its diversity and bulk continually increased to meet the needs of growing populations. Despite this, it is still argued sometimes that the correlation between happiness and technology can be inverse, due to an increasingly cold, mechanistic world fitting badly with minds and bodies that are claimed to be geared to gathering and hunting. However, in addition to the benefits just listed,

SUPERINTELLIGENCE AND WORLD-VIEWS

progress led to increased scope for more expressive or creative activities. Not only philosophy is the 'child of idleness'!

Some outputs from much of this development activity can be thought of as man-made imitations, improvements or replacements of human body parts and other natural 'machines' and human 'operations' in the widest sense. In many places 'living' has almost been re-invented in recent decades, not only due to pop-up toasters, electric drills, audio tape, and mobile phones, but we also have the WWW, mobile computing and Watson now, and Concorde has come and gone. Aircraft that can use conventional runways can do more than 2,000mph, and some aeroplanes can fly at over 7,000 mph. The Chinese supercomputer, Tianhe-2, can calculate at over 33 petaflops per second on the Linpack Scale, and a computer which could forecast weather for two full weeks ahead is predicted for about 2030. And so on....These advances form the backdrop for our contemplation of man-made improvement of reasoning, learning, and other cognitive capabilities of the sorts envisaged for SIs.

What is the flipside to all this more recent progress, and is there anything to make us cautious about any movement towards SIs? Consider a detail of our rapidly changing, technologically tied-up, world, namely, the 'degree of freedom' which has been provided due to the fact that you can answer a personal phone-call from a mobile almost anywhere in the world. A *Scientific American* 'Recommended' feature in Aug 2014 looked at: *"four books this month (that) investigate how our increasing dependence on mobile devices, social media, and the online world is changing the way we feel and think"*. One price we pay for being artificially networked so much is that we experience: *"a deficit of silence and solitude"*. You can't 'get away from the phone' as easily as was possible when I started out on my career. Furthermore, keeping an orderly mind is actually more challenging due to the great tsunami of information streaming from technology these days. People are also sometimes: *"merely dazzled by phenomena"*, to use John Newman's phrase[5], so that from the voluminous and thinly spread information, we get, there is a: *"quick burst of satisfaction that actually quells our inner hunger for understanding rather than seeding it"*. Privacy concerns are also raised amid the undoubted commercial and enabling promise.

This ubiquitous upheaval and challenge presented by innovation could be due at least in part to a technological world-view that many influential people hold, which includes the, perhaps tacit, goal of 'Technology Saturation'. But it could be said that technology is already too widespread. For example, it's too easy for us to check in to one's work when we're supposed to be 'away' and unplugged, and so staying connected prevents us from getting a break from work. We could, of course, take a vacation in some remote area with no telecommunications access. An alternative rumoured to be offered as a service [sic] by some suppliers, is of specially arranged 'mobile holidays' of, say, a weekend, allowing subscribers to be unplugged and freed from such accessibility via mobile phone. However, withdrawal symptoms might be expected to be manifested among smart phone owners. In the US, over fifty per cent of these people allegedly check their phones at least once every hour. Nevertheless, short respites from e-mail, Facebook and/or Google search would surely be welcomed by a lot of people. Technology has proved to be useful as it has led to spectacular advances, but this does not change the fact that the increased production of less than essential goods and services has had some of those less than essential, and perhaps less than desirable, effects on the quality of life.

On a different tack, on some accounts, another fall-out from automation and powerful control devices seems to be *increasing over-self-confidence*, and this is an important consideration for any contemplation of movement towards SIs, and our evaluation of suggested End scenarios like the Omega Point. For example, claims are sometimes being made of mankind getting close to a stage of development where we are the authors of our own destiny, and we have a modern Tower of Babel situation. Life independent of any guidance from 'something greater' is contemplated. As we saw in Chapter 6, technology and science are looked to by people like David Deutsch as the most promising potential means of salvation on an overcrowded planet and as a conduit for fulfilment for us humans. With improving computers and other devices, they put forward the possibilities of forming 'bespoke virtual reality', for example. But people can't just turn their backs on known, sensed and experienced reality – the real world. And environmental responsibilities and even transcendental built-ins, for example, can't be simply forgotten about. However domination of everything else by humans using technology, can be seen as a tacit goal, and narcissistic boosting of human egos could be dangerous. Of particular interest here is the suggestion that creating human-like intelligence that can be seen as some sort of ultimate human achievement is the epitome of this conceit. Even more arrogant is the assumption that a human could create a superior being 'in his/her own image', and that such a being would be profoundly interested in us and our values and goals.

SUPERINTELLIGENCE AND WORLD-VIEWS

The French philosopher Jacques Ellul, referred to[6] the collection of technical accomplishments at the middle of the 20th century as a new 'milieu' which had replaced 'nature'. He used a key phrase to characterise all technological change: *"the collision between spontaneous activities and technique is catastrophic for the spontaneous activities"*. He describes a civilization increasingly dominated by technique as a tragedy. By 'technique', he meant more than technology. In the foreword to Ellul's book Robert Merton writes that it is: *"any complex of standardized means for attaining a predetermined result"*. For example, a totalitarian political regime would qualify, but the concept is also reasonably applicable for our considerations here. New technology is good for some things, but not everything, in life. We wouldn't want to miss out on the experiences of Heaney or Wordsworth or even our farmer or cottager of Chapter 4 by being over-committed to it. Ellul observed that: *"aims of technology, which were clear enough a century and a half ago, have gradually disappeared from view"*. This has important impacts on mankind. Even when Ellul wrote in the 1960s, the 'Technology Saturation' goal was leading to the fade-out of the view of technology as being purely a pragmatic enabler. He wrote: *"Our civilisation is first and foremost a civilisation of means"* and *"in the reality of modern life, the means, it would seem, are more important than the ends"*. He said that technology had an artificial life of its own and that it was self-determining, and so resistance is needed if we're not to be swamped by either the, often facile, possibilities or the constraints of technological systems. We can't just continue without much 'thought for the morrow'. The great variety and qualities of our culture could come under threat unless the thrust for technological advance can be controlled. A vision of the horror of universal drab sameness would loom with culture being confined to what technology allows and offers. The prospect of Emily-like and other creative artificial agents at W^*25 level becoming available, equipped with a plentiful supply of genius-tricks, might alleviate this modestly, but consideration of quick rewards and benefits would be likely to dominate any thoughts about any less tangible and longer-term costs.

Being Careful. We'll be looking further at the realistic future for AI in Chapters 13 and 14, but I want to flag just one or two obvious related technological threats that arise in addition to those listed above. Technology-dependency is not necessarily good *per se*. Think of power generation and other such utilities that are essential to techno-life and to the economy. The well-worn warning is as pertinent as it ever was – we don't want to get to the point where, if we lost our telephones, televisions, internet, cars and trains, our whole culture would collapse disastrously and our very existence would be threatened. Moderation of the appetite for gadgets and tools is something that comes to mind as a desirable objective in this context. Furthermore, man-made tragedies such as that at Bhopal, and the effect of natural disaster on technology, such as that experienced at the Fukushima 1 Nuclear Power Plant when it was hit by a tsunami triggered by an earthquake, are salutary. How much more challenges would powerful and free SIs present? Building a free-ranging and free-willed Frankenstein's monster of an SI would not stand out as a good idea, and some of the modern droppers-out we thought about in Chapter 4 would be inspired negatively if such a reality loomed.

Technology is developed in social, environmental, economic and political contexts, and within other narratives. It shouldn't be seen as some sort of 'bigger than us' phenomenon that will lead certainly to some Utopia. And problems not fitting into the technological model can't be simply ignored. An undesirable, probably unacceptable, spiritual void could be the ultimate result of thoughtless 'advancement'. Having and using a reasonably sound and near-complete cosmos-view under human control would mean that nature, along with values, culture and other aspects of civilization could be taken into account, avoiding crippling entanglement in technology and possible destruction of that nature and culture.

So, when 'advancement' towards something like SI is proposed it should be vetted carefully. Tools and methods that are sustainable, and people-friendly, environment-friendly and culture-friendly should be prioritised, and maintenance of the best established values is a must. This leads to some familiar debates – for example, what is the level of responsibility of those who innovate technologically or propose scientific advancement? Things like integrity, honesty and high standards of skill are certainly obligatory – but can the inventor of a spade or a wheel be blamed for the dangerous misuse of these? So what if it is SIs we're talking about rather than spades? Collective responsibility is more appropriate here as we try to predict and avoid likely problems.

References

1. Midgley, M., "All too human", Guardian.co.uk, 19 Dec 2008.
2. Midgley, M., Evolution as a Religion, University Paperbacks, Methuen, 1985.
3. Gee, H., *In Search of Deep Time—Beyond the Fossil Record to a New History of Life*, Free Press (Simon Schuster), 1999.
4. Vidal, C. Metaphilosophical Criteria for Worldview Comparison, *Metaphilosophy* 43 (3), 2012 http://homepages.vub.ac.be/~clvidal/writings/Vidal-Metaphilosophical-Criteria.pdf
5. Newman, J., *On the Scope and Nature of University Education*, Everyman's Library, 1915.
6. Ellul, J., *The Technological Society*, John Wilkinson, trans. New York: Knopf, A. A., 1964.

Chapter 11

Values and Memes.

Does a dog have any *values*? Does Watson? Human agents do. They tend to be constrained or motivated, directly and powerfully, by the values they adhere to most passionately. Their values dictate any 'principles' or standards they set up for their lives, and to some extent how they react to and relate to people and the everyday world. In the world-view schema, A-schema, suggested by Apostel's group that we looked at in Chapter 3, for example, the *values* functional component is seen as being necessary for distinguishing good from bad, often by means of rules, and for setting objectives for behaviour, and giving direction to our activities. To get some idea of what values might be imagined for an SI, and of the difficulties posed by choosing from the contender values that humans often have, we will consider again our 'conversation' from Chapter 1 between those two, very much in the future, versions of W*.

To date no-one has *shown* us any AI approaching the self-awareness and reflective capabilities of W*22. W*1 (taken here to be, roughly speaking, that combination of Watson, Deep Blue, etc, and the superstructure shown in Fig 2.1) has amazingly fast access to a lot more immediately available 'knowledge' than most, if not all humans. Missing capabilities are capacities for self-awareness and the ability to reflect on its own experiences, but also things like values, visions and aspirations that would lead it to a stage where it could contemplate 'pulling the wool over the eyes' of humans. However, we can use the W*21/W*22 scenario of Chapter 1 here for two purposes – as a *sort of* motivating example and as a potential test case for the study of some aspects of the possibility of having values in machines, and how those values would be acquired and developed. We'll look at that sort of thing shortly, but I want to fit it

SUPERINTELLIGENCE AND WORLD-VIEWS

in with what we've been considering about some component world-views in Chapters 9 and 10. From a different angle, the more functionally descriptive structure and pattern of the A-schema, might, in fact, be at least partially inheritable by those respective component views. World-views will include *values* and *actions*, and also ideally: a *description* and an *explanation* of the world, as well as a *prediction* of what the future holds, an *epistemology* – what is true and what is false. Here I want to look primarily at the values functional unit, with a little bit on the actions. There will be some more on description and epistemological issues in Chapter 12.

Values. You probably won't have been able to pick out a lot about the cosmos-views and world-views of the two agents, W*21 and W*22, from that short dialogue in Chapter 1. However, there are some hints about attitudes and we *can* get some clues about their values and possibly their goals from it. A start could be made by looking at their possible reasons for decisions, such as that humans won't spot flaws when given a 'proof', and motivations, like self-preservation when shut-down is threatened, and linking these to possible values. More importantly, are there any particular values we should *try* to ensure that W*s, and the like, should be given, that they can't sneakily ignore or change through some intervening technological advance?

Are there things that have 'true value' rather than simply being desired? This question has no simple answer. In the old world described by Bloom which we looked at briefly in Chapter 6, the old 'absolutes' made choosing values much simpler for the good folk who adopted the regular, given world-view or cosmos-view. Adam Smith[1], in 1759, listed some basic forms of attitudes and behaviour, which many in that old world would have included on their lists of virtues – that is, the human attributes they valued most. Smith wrote positively on *empathy*, which involves taking into account the feelings of others when we're deciding and acting. He also advocated *justice* between people and groups; *self-government* where the 'passions' or 'carnal appetites' are well controlled; and, *obligation-honouring* or doing your duty. But these aren't all appropriate for, and even applicable to machines, and even in the human world, how do these four values square with the values manifested by Cobbett, Livingstone, Koestler, and the others we've looked at? And should Ava and Sonny have been given some well-considered values before they set out alone? Where would they come from?

There do seem to be certain positions that can be taken on values that are seen to be nearly self-evident, and which appear as effective *de facto* defaults for decent humans. Some might even view them as FAPP *absolutes*. For example, it is almost universally accepted that all human life is precious, and that many moral laws and rules are indispensable. Some values are even being considered widely as being innate to the human psyche. Basic moral constraints – prohibition of stealing, killing humans, lying, and adultery can be spotted across diverse societies. For example, they are included in the biblical 'Ten Commandments'. Universal morality considerations such as looking after one's family tend to be based on an 'obvious' understanding of the human condition. 'Do unto others as you would have them do unto you' is the very well accepted 'golden rule'. That the state of pleasure is preferable to that of pain seems to be accepted generally as being effectively universal. Tolerance of strangers and 'other' cultures is perhaps not so widely sought. On the other hand, the celebration of beauty and love as important values puts them on many lists, which include other fundamentally important values, like courage, that more or less everyone admires.

There are many 'values' that are not near-absolute. They have meanings that are seen to be conspicuously variable, and it is well accepted that it is foolish to try to put final, tight meanings on them. Consider Adam Smith's list[1] of values above, or a list like compassion, truthfulness, fairness, freedom, unity, tolerance, responsibility and respect for life. Lists of values overlapping with this latter list of eight values and with each other appear regularly in the literature. The list just given is a selection from one particular system[2] that purports to judge on vocational profiles of testees who choose those they favour from a total list of sixty. The question for us is: are such lists of any value themselves? In particular: do they input significantly to world-views and cosmos-views? And which of them would be suitable for imposing on, or simply suggesting to, machines?

Take the influential theory of personal values designed by Milton Rokeach[3], again intended mainly for use in vocational profiling. According to this scheme, people each have thousands of attitudes towards thousands of specific things. Those using values from the list for vocational profiling are asked to reflect on their meanings and on whether each particular value is something they want to influence decisions in their working lives. Rokeach[3] saw an individual's values as being 'modes of behaviour' they would like to adopt, or a 'state of existence' they would like to end up with. So, for example, W*22's proposed deception tells us something about its values. Rokeach published lists of values for use in various practical applications. His *Terminal Values* are concerned with desirable goals

SUPERINTELLIGENCE AND WORLD-VIEWS

– things that an individual would like to achieve in life. These are outcomes someone might want to target, or might characterize a person's ideal world. His 18 suggested Terminal Values were: freedom, happiness, wisdom, self-respect, mature love, a sense of accomplishment, true friendship, inner harmony, family security, a world at peace, equality, an exciting life, a comfortable life, salvation, social recognition, national security, a world of beauty and pleasure.

Rokeach produced a value system called the Rokeach Value Survey, which is really a classification scheme for values, in his 1973 book[4]. He includes *Instrumental Values* in his scheme. These are the 'modes of behaviour', or means of achieving Terminal Values. There are 18 of these again: cheerfulness, ambition, love, cleanliness, self-control, capability, courage, politeness, honesty, imagination, independence, intellect, broad-mindedness, logic, obedience, helpfulness, responsibility and forgiveness. It is interesting to again imagine applying these to Livingstone, or Koestler or some of the others we've met in our 'case studies'. What about Cobbett and the subsistence farmer, or the Brazilian tribesmen, or W*22 or W*101?

So much for a *sort of* bottom-up approach to the idea of values. From a top-down perspective, which might help us home in further to some characteristics that SIs should or could have, values are often said to offer humans roots that can be used as anchors in the storms of everyday life, as suggested by the Latin root of the word – *valeo*, to be strong. The term *axiology* meaning the study of value has become a respected topic of scholarly inquiry. Since at least Plato's time, philosophers and others have tried to pinpoint exactly what values should be observed in an ideal world. They have been described in many ways – including lifestyle priorities, practical philosophy of life, outlook, deeply held principles, ideals, or standards. Obviously culture and STEEP and other factors can affect relatively local values and even what are considered more absolute values. Teachers, friends, and especially parents influence values early in life and they tend to be persistent – remaining fairly stable over time.

Values in general are things a person will tend to stick to very consistently. Some indicate a lot about what that person is known for, and what they can be expected to do or decide, even if this results in something that is unpopular and difficult. Others are interested in his/her primary motivating drives and his/her significant resource consumers. What modes of conduct and beliefs seem to be foundational or ultimately important in their cosmos-views? Values affect cognition, emotion, and behaviour, and often there might be trading-off between values from a personal 'value list'. Moreover, it is likely that many items on the list will be shared by many other individuals and groups. If we assume those terms somehow cover much of the most important concepts of values they could be collected together and be put into a *sort of* reference vector, and this could be sorted as required by any particular group, individual or future W*. In 1852 it was written of David Livingstone[5] that he wasn't driven by *"...mere curiosity; or the love of adventure, or the thirst for applause, or by any other object, however laudable in itself, less than his avowed one as a messenger of Christian love from the Churches"*. Some items from the lists above are reflected in this statement. It is interesting to ask how people like Livingstone obtain their values, and that is something, which I leave for the reader to complete as an exercise.

Nearly all of the exemplar words or terms in the lists of values above tend to mean something that's just a little bit different to different people. Indeed, in the opinion of some people, all values are relative or subjective. A group of doctors may come to an agreed understanding, a consensus, which may be informed to a degree by science, on how to determine when life has ended, or what the, perhaps somewhat fuzzy, 'well-being level' target is for a given population, such as a geographic region. These might change from time to time, as new facts become available. In this case scientific facts or insights are pertinent. Scientists and engineers, like medics, also make relative value statements like this, and we have to acknowledge that it might prove impossible to identify one 'right' set of values that appeal to everyone, or, even harder, a set that we could profitably put into general use with any future SIs.

However, it is intriguing to note, as we did at the outset, that there *is* a set of values that might be fixed for everyone – in other words, they might be absolute – like Trust and Justice. Some of these could provide us with initial contenders for an SI. Even if a term is not genuinely absolute, it is possible that it can still be said to be *ideal*, in a sense. It could be an unattainable limit to which any real agent can at best try to approach asymptotically. Practical 'well-being', which we linked with 'happiness' in Chapter 2, is an example of a value that might mean something different to, say, a hospital patient and what it means to an injured soldier in a battlefield trench. In Chapter 2 we also saw that the 'why' might outweigh the 'how' to give happiness even when in dire straits. We find it very hard to identify anything at all that can meet *all* the criteria for *all* the classifying attributes for happiness perfectly for humans.

SUPERINTELLIGENCE AND WORLD-VIEWS

Values for, say, Ava or W*22 are at least equally hard to pin down precisely, and values for SIs' world-views remain for grabs. Like humans they would need to be aware of any absolute values and know what they mean, especially to humans, from experience or by tuition or by revelation, which leads us directly to some of the issues discussed in Chapters 7 and 8, and machines would probably always have to be given some of these. Knowledge-seeking methods, such as those used in science in its broadest sense, which could very feasibly be used for exploration by artificial agents, as we considered in, eg, Chapters 5 and 10, can inform relative values, giving crisper definitions to relatively fuzzy concepts that by consensus are already accepted as being positive. Machines and their users would also like the value of something to always be greater than the cost incurred, but often these have to be consciously traded off against each other. This pragmatic approach could be useful in practice for humans, and maybe even for in some situations, but *per se* it does not help us or the SIs to understand absolute values fully.

One researcher, Bond[6], gives a simple taxonomy of levels of motivation for valuation: The first level is *simple inclination*, that the agent just felt like it. A little more complex is such a felt *pro/con gut feeling* plus a belief that some act offers satisfaction. A third motivation is where one is motivated *out of a particular emotion* that explains the desire one has for the end that is targeted. To have any of these a W* would have to have feelings, and presumably this is the sort of motivation behind W*22's utterances and suggestions. We humans can generate such values by answering the questions 'why do this?' or 'why choose that?' Possible answers include basic usefulness; the promise of pleasure; the aim of avoiding harm; the constraints of moral correctness; the pull of loyalty; and nobility of character (eg 'it would not be worthy of me to do this'). In the case of the W*21 and W*22 story, prevention of harm to *the agents themselves* was probably their motivation.

It is hard to see that W*22 or any other artefact 'on the horizon' can have such values unless they are built-in simply as playback versions. It is not altogether clear how effectively the values could be incorporated in the future machines by their (perhaps ultimately mechanical) designers. Maybe letting W*s free to *"roam the country side"* as Turing once suggested, and as Ava could have done, would be pushing our luck. Turing warns: *"the danger to the ordinary citizen would be serious"* [7]. This would, of necessity, greatly delay the date when we can expect them to have their own cosmos-views. In the meantime, however, values that do come from our groups of doctors, scientists or engineers as above, who arrive at an agreed understanding, a consensus, on how to determine or define something, can be given to a W*. Or the knowledge transfer could be done much more efficiently, as the successful use of mechanisms available to the current Watson suggests. Moreover, most original discovery methods used by human experts to get Watson's source knowledge were empirically based, founded on observation and fitting with other facts or insights. Some of these sources of input would be available to the W*s. and perhaps successive generations of W*s can improve on them. But if the full 'SI narrative' ever materialises, at W*25 level and beyond, the judges of the quality of any development of values are likely to be SIs.

Built-ins and Good Naïveté. I used to have a dog, Flojo by name, which did not conspicuously exhibit any of Adam Smith's four values, and which I would have been very reluctant to train as a sheepdog. She was the hardest dog to house-train that I ever had the pleasure of knowing. Even for a wilful terrier she had an exceptionally resolute mind of her own. So, if I needed some help to move sheep around a field or enclosures I would probably not try to train Flojo for this, but I would choose to start out with a well-bred sheepdog puppy, say a Border Collie pup, which would come complete with a predisposition to round up and a willingness to learn. It is possible to see young puppies of this breed torment the lives out of the poor ducks as they instinctively 'herd' them in a farmyard. *Innateness* is demonstrated here; in this case it is that predisposition to herd, which is the result of careful breeding over many generations, and Flojo did not have it.

Humans come equipped with faculties dedicated to rudimentary reasoning about space, number, probability, artificial and natural physical objects, living things, and other categories. Babies start out with plenty of neurons to support their mental advance during 'critical periods', especially in their first year, and the brain expands rapidly in size to accommodate subsequent development. Certain skills, tastes or abilities appear according to a rough schedule, and they are said to be native. For example, newborn babies have built-in linguistic propensities, even though they don't come ready-equipped with verbal language on which to base communication with others. The predispositions they do demonstrate are usually shared by most babies to some extent. From the first few weeks after birth infants manifest various movements or reflexes. Some are spontaneous, a sort of baby babble of limited actions, and others are responses to particular triggers, such as the reflex to suck, where the baby sucks a finger, and reflexive grasping, where the fingers close given the right stimulus, and the infant can sometimes hold things quite tightly. Other examples are where a baby appears to try to take steps when held upright with feet just touching the floor, and when the baby 'jumps' when it hears a sudden loud noise. I'm sure everyone's seen at least some of these.

One of the prominent aspects of knowledge arising from recent baby-study is that it has been demonstrated that certain capabilities appear at very early ages. There are clear indications that we are all kitted out very early in life with some naïve physics, naïve biology, and naïve mathematics. Parents and grandparents know well that newborn babies tend to have an inherent tendency to be attracted to certain things. The classic example is that 'eyes' of any sort clearly attract a baby's attention. Prominent features of an object often hold the attention of infants as young as one month. More generally, in the visual realm they can detect changes in brightness, focussing on straight edges or prominent curves. Studies have shown that newborns recognise their mothers' faces two weeks after birth. Even at age zero they can distinguish between stationary objects and moving objects, and they can track moving objects in their visual fields. The longest interest times are associated with the most unexpected events, and sometimes such interest is detected when things that are in fact physically impossible are presented, like certain occlusions or solidity violations.

Experiments have led psychologists to conclude that we are born with a *naïve physics* in place – an innate ability to understand certain aspects of the physical world. This basic physics, duly refined and developed through experience, would probably be very valuable for hunter-gatherer families, say. For them good capabilities of this kind were or are needed, from early in life, to enable a young person to make decisions about moving things, for example, quickly and accurately. Naïve physics manifests itself in commonly understood, intuitive, or everyday observed rules of nature such as: a dropped object does not usually stop suddenly in mid-air; a box 'inside' another is also inside a box that holds the first (and this containment has certain size implications); and two events happen at the same time or they don't. Naïve physics usually harmonises nicely with daily experience and observations, but sometimes the naïve view turns out to be wrong. Optical illusions come to mind, and sometimes our naïve dynamics go counter to Newton's laws of motion. For example, what way would a ball projected through a shortish twisted hose-pipe lying on a table emerge? People respond surprisingly often by saying that it would continue in a somewhat spiralled trajectory, or at least trace a curve of some sort. Fortunately such wrong predictions do not lead to disaster as the naïve 'laws' often 'work' FAPP because of the relatively small scales and limited scope we deal with every day.

There is a similar claim that we all come equipped with another received framework of perception and understanding, a sort of *naïve biology*, this time ready for use when we start to consider the differences between inanimate things and living things. Populations everywhere classify the things we've labelled as 'living' into species-like groups, which seem obvious, and are often shared across cultures. The linguist, George Lakof, wrote about this [8]. For example, the members of species interbreed amongst themselves, and the individuals in any particular class have some distinctive basic essence, responsible for their shared classification, such as appearance or patterns of behaviour. These classes are assemblies of individuals usually arranged in a hierarchy of sophistication or complexity. They are often apparently based on needs FAPP, or on intuitively or culturally pleasing or natural concepts. Importance to activities in the everyday lives of the group members is a big influencing factor. People in all cultures seem to, more or less universally, sub-divide animals and plants roughly into classes that expert biologists also regard as generic species. They usually have a liking for some environment. The groups compose a generalisation hierarchy, like a lion *is-a* big cat *is-a* meat eater. The tendency of humans to build up naïve taxonomies such as this to organize 'kinds' of plants, minerals and animals in groups seems to be innate, and some people believe that there are natural kinds in the world, underlying any linguistic analysis of the 'essence' of the kinds, and underlying folk classification. Naïve biology is based on the resulting taxonomies. Incidentally, communication via language appears to have a strong relation with such classification. For example, quantifications like '*For all* cuckoos' and '*There existed* a foster-mother', are widely used, and even the presence of a few atypical examples doesn't put us off.

Number usage by humans has been explained from the adaptation to FAPP requirements or advantages. There were likely to have been advantages of basic reasoning about numbers in, say, Africa for hunter gatherers, say near Laetoli or Afar, where there were finds of hominid remains. 'I had three babies when I got here, I now have two. I'll have to look for baby three'? Although the Platonic

SUPERINTELLIGENCE AND WORLD-VIEWS

conception of number respects the mind-independence of, say, the number 'two', many mathematicians and philosophers say that numbers, orders and shapes, while existing in abstract space have real properties that can be used for baby management, for example. Further inspection of this naïve 'given' takes us into the wider domain of *naïve mathematics*. A baby of a few months old is stirred when things 'do not compute', at least at the level of very simple arithmetic. Suppose an item such as a toy is shown to it, and then covered up by a screen, and then another toy is clearly 'added' behind the screen. The baby is subsequently surprised if there is only one item when the cover is removed. And most young children enjoy games involving numbers and the other fundamental mathematical objects or concepts. In addition to the *numbers*, used from an early age to count, label, and measure, these concepts include *orders:* arrangements that make some sort of sense among the separate elements of some collection, and *shapes*: spatial properties of the physical volume or area taken up by particular entities. Moreover, while some would say that statistical and probabilistic thinking is an artefact of relatively advanced civilisation, there would always have been a need for reasoning under uncertainty, even in the most primitive lives. It has to be pointed out, though, that it has been argued that intuitive probability, for example, is very shaky and often inaccurate even in present civilised communities, as Amos Tversky and Daniel Kahneman[9] convincingly demonstrated over the years. Even intellectually sophisticated people are not naturally adept at statistical thinking, and intuitive feelings for 'probabilities' are often at odds with the laws of mathematical probability. We have, for example, a tendency to 'anchor' the probability of an event to how easy it is it can to think of examples.

Innate behaviours in all species could be triggered by features of the environment or from some sorts of stimulus response and nervous reactions. The responses to sensory and nervous stimuli may have been developed to give the holders a built-in advantage over individuals that don't react so well to circumstances. Obvious examples are a gazelle's adrenaline rush and flight on hearing a loud noise or the way fledglings of various kinds 'freeze' when they see the silhouette of a peregrine. Think also of the cuckoo's parasitical behaviour as it lays eggs in the nest of chosen foster parents as a demonstration of response to sensory stimuli. Rich proximate patterns are displayed in the use of navigation skills of homing pigeons based on, say, on the fly (literally) visual and magnetic cues. The migratory behaviour of terns as they travel huge distances from top to bottom of the globe, combines innate and learned knowledge. Web building, pecking orders, territory marking, and communication and altruism in ants and bees, also leave some scope for learned adjustments and improvements to received capabilities.

In the 1990's investigators pushed *ethology*, a sub-topic of zoology concerned with the scientific and objective study of animal behaviour, and it emerged as a formal discipline, looking at, say, how geese, bees and gulls, for example, behaved. Lovely pictures can be found on the Internet of one of these pioneers of ethology, Konrad Lorenz, being followed by a line of new-born goslings, probably uttering: *"Mum! Mum!"* in goose language. Lorenz was the first moving object they saw just after they hatched, and their imprinting led them to 'assume', and thus in a sense learn by habituation, that he was their mother.

There are possible hints in those patterns of human development for any would-be designers of smart machines like W*22. Some built-ins could be given to the novice agent to equip it with naïve predispositions to reason in certain ways, and to perform particular patterns of actions, as we'll see just below. Such action patterns belong to the 'action' functional unit of that A-schema framework developed by Apostel, et al. Clearly, not all patterns of action should be built-in, being left for later acquisition through learning by imitating others, for example. However, some of these developments could be triggered by built-ins. This leads naturally to the concept of cultural development – learning by shared practice, common experience and collective reflection for successful and profitable development within communities. This involves ways of co-operating and sometimes ways of handling conflicts.

Memes and Culture. Nods and winks, along with certain mystic and ceremonial practices, a wealth of religious rituals, and a variety of other patterns of action are referred to by some people as *memes*. They can be observed in different modes of religious worship, and in some sorts of national events, such as the Remembrance Day traditions in the UK, and fit nicely into the action part of that A-schema. Let us look for a moment at what I take as a very common, and probably the most conspicuous, aspect of memes – namely that of an agent *imitating* how some other agent does, or simply describes, something or other. This is potentially a feature that SI designers could hope to make use of. It's fundamentally about learning by copying. The word 'meme' itself is actually related to the Greek word *mimos*– to mime. If we look at each of the examples above in turn, we see they have certain features in common.

SUPERINTELLIGENCE AND WORLD-VIEWS

They can all be fairly readily and (more or less) universally learned and used, with some small variation, by all of the individuals within some groups, and there can be accumulations of, sometimes subtle enough, modifications over time.

It can be readily appreciated how the idea of looking at some things in this particular way, and calling them 'memes', was inspired by a very limited analogy with genes. The underpinning idea behind the concept of memes, which was first presented by Richard Dawkins[10], is to substitute: *"ideas, catch-phrases, clothes-fashions, ways of making pots or building arches"* for genes. Some people say we need to know in detail how memes are represented and instantiated in the brain before we develop a full-blooded theory of memes in line with that of genes. There is also a weakness due to their super-ubiquity and their ultra-wide scope noted above. However, that said, as a broad conceptual framework for looking at the propagation of ideas, culture, and ultimately world-views, the notion may have something to commend it.

Consider as an illustration the use of the notion of memes in one of its most evident applications – the study of *culture*. The term 'meme' is indeed used widely simply as a label for *a fundamental unit of cultural inheritance*. Now, although there seems to be some evidence of physiological, neural substrates of culture, culture itself is considered to be beyond either the thought or the biology of any individual. Susan Blackmore[11] is one of those who have prominently described culture in meme terms – looking at culture as a meme's way of replicating itself. And indeed there are aspects of this that could be important in the world-views of those individuals within that culture. Consider family traditions regarding the details of bringing up children, or the cooking of goose rather than turkey at Christmas.

Some researchers have pointed to further deficiencies in the gene-parallel stance. There is no acquisition of the modifications from parents by offspring, over an extensive family tree that we would be looking for in true cultural evolution. The knowledge may not necessarily be shared by all, or most, of the individuals within the group. There are other good reasons to question this evolutionary, memetic, take on world-view development in particular. In my view what is most telling against it is the fact that people, and possibly some very smart machines, are essential for the generation of the sorts of ideas we've been talking about. They accept or reject them through the exercise of their powers as *intelligent agents* – as they deliberately *think* and consider such things as values, means and ends, costs and benefits, and purpose, objective, functions and scope. This makes trying to exploit an analogy between the spread of ideas and evolution's natural selection very suspect. Consider the fundamental stipulation of natural selection that *blind physical evolution eliminates the need for conscious, purposeful thought*. Also on the negative side, when the multi-modular human mind is studied cognitively, it is observed that, unlike genetic replication, high fidelity transmission of cultural information is the exception rather than the rule. Mutation rates will be high and inaccuracies will accumulate quickly. The fine-grained epistemic exhaust is extensive. During communication messages are generated that could be cripplingly widely-varied. So complex models are required for interactions, trade-offs and balances between various factors. The evolution of genes depends on balances, often fine enough, between selective factors and mutation rates. The lack of evident equivalents of such features adds to the clear perception that considerable work is still needed for the interesting analogy between meme and gene to become anything more than that.

The advance of societies does, of course, produce much 'epistemic exhaust' and the innovators pushing things forward are not always high-powered masterminds. For example, genius-tricks aren't always needed: *"...most of the cultural spread that goes on is not brilliant, new, out of the box thinking. It's 'infectious repetitis'"*[12]. Still, parallels between propagation methods in culture and genetics are not totally without appeal, and again aspects of the meme concept could be useful when considering the possible development of the world-views of SIs. Perhaps W*25 with advanced modules, maybe including some genius-tricks, could 'pass them on', after any improvements that it might dream up, in some similar way? Further insights that would be useful for intelligent machine techniques might also accrue if the analogies with culture and genetics are clarified. At present memetics promises to provide a *sort of* philosophical shuttering, despite not being a useful scientific object.

References

1. Smith, A,.*The Theory of Moral Sentiments*, (1759).
2. http://www.emotionalcompetency.com/values.htm
3. Rokeach, M., Long-range experimental modification of values, attitudes, and behaviour, p 453–459, *American Psychologist*, 1971.
4. Rokeach, M., *The Nature of Human Values*, The Free Press, 1973.
5. Blaikie, W. G., *The personal life of David Livingstone*, John Murray London 1880. A quotation: Thomas Jefferson to Peter Carr, August 10, 1787.

6. Bond, E. J., *Reason and Value*, Cambridge University Press, 1983.
7. Turing, A. M., Intelligent Machinery, in *Cybernetics*, Butterworths 1968, eds Evans, C. R., Robertson, A. D. J., from *NPL Report*, HMSO, 1948.
8. Lakof, G., *Women, Fire and Dangerous Things*, University of Chigago Press, 1987.
9. Tversky, A., Kahneman, D., *Judgment Under Uncertainty*, Cambridge University Press, 1982.
10. Dawkins, R., The Selfish Gene, Oxford University Press, 1976.
11. Blackmore, S., *The Meme Machine*, Oxford University Press, 2000.
12. Dennett, D. C., *TED talk – Dangerous memes*. www.TED.com. 2002.

Chapter 12

Knowledge – Immature, Enigmatic and Partial.

We now look at the *description* and *knowledge* aspects of the A-schema we're using, as suggested by folk at the Center Leo Apostel in Belgium. Our cosmos-views contain some broad concepts of how the world we live in is constituted, how it functions, and how it is structured, and we need to be able to build reliable models of this. There is also an obvious link-up with the prediction and explanation aspects. Now we live in privileged times when the science and mathematics under-pinning calculations such as those needed to get a space station to work reliably, effectively and efficiently, and even those in more uncertain medical reasoning, weather forecasting or economic analysis, appear to be much better for accounting for the past and discerning the future and knowing what fate might hold, than more ancient ways of doing this, such as by studying the entrails of animals or the trajectory of some star across our sky. The transport systems to get from Belfast to Nanjing in a reasonable number of hours would probably not be available if we had not moved on from such primitive prediction devices. Discoveries leading to the state of the art in technology have leaned heavily on a formal method called *induction*, which is pretty good, FAPP, at getting the predictions right for things such as components of aircraft, but it is not so good for medical purposes – probably because biological entities are not designed, controlled and adjusted by engineers as artefacts such as planes are.

Induction does not, of course, exclude biological entities, and it is just as important in the study of the natural world, especially where populations are concerned, as it is in that of man-made systems. Imagine an ornithologist observing a large group of swans on a large, newly discovered island over a long period – one swan

SUPERINTELLIGENCE AND WORLD-VIEWS

after another. Suppose the ornithologist obtains an observation sequence in which only white swans are ever found. The principle behind induction can be seen in action here, as much as in the study of engineering systems. If an 'all white' sequence is recorded, you could arrive at a 'working generalization', viz *all swans are white*. It could be possible that there are some black swans, which are never observed, but even then the generalization would always give good predictions. If eventually a black swan *is* found, then we would like the ornithologist to arrive at the conclusion 'not all swans are white'. This could of course be refined over time – for example: 'In certain parts of the island all swans are white'.

There are weaknesses in induction, leading to a fairly complex logical problem, called Hempel's Paradox, about what would 'confirm' the ornithologist's swan rule, for example. David Deutsch[1] has illustrated a more obvious weakness. Up until the turn of the millennium, Deutsch always experienced dates with years starting with '19'. However, although this had happened thousands of times, and he'd never encountered dates with years that start with a '20', the: *"...explanatory theories"* he had: *"led him to expect them..."*. Incidentally, Henry Gee who was quoted in Chapter 10 would probably say that a generalisation that allowed us to expect a '19' would possibly have been better founded than some of the generalizations we see from understandings of the history of Earth. On the other hand, designers of aircraft, and ultimately passengers taking a trip to Nanjing from Belfast are prepared to depend, in a rather basic way, on knowledge gained by induction.

To get inductive knowledge we start from particular facts, and produce general rules. There are other ways of building up knowledge bases. Deductive knowledge is gained when we go from generals to particulars – *Dogs bark; Fido is a dog; so Fido barks*. Some 'general working knowledge' is acquired directly from the senses, or from some authority on the matter in question, or from some gut feeling or intuition that a proposition under consideration is true. Basic facts such as those of naïve physics we looked at in Chapter 11, eg that the whole is greater than any one of its parts, give us a start in this and they are taken as granted by most people. 'Knowledge' can be added as the truth of some new proposition is established as a chain of reasoning involving connections between facts and rules that are taken to be true, *ab initio*.

Human Knowledge. There is a sense in which our knowledge is *immature*. Consider the way that a child understands his/her world. I was told as a child that storks had a big role to play in baby production. Now, this very innocent belief was at least matched in immaturity, in the present usage of that word, by those of even distinguished scientists of their day, such as Jan Baptista van Helmont not so very long ago. Van Helmont was a well-known Flemish physician and alchemist, respected by people like Robert Boyle. Surprising to most people is the fact that, as recently as in the seventeenth century, the time of Newton, van Helmont 'verified' a belief in Spontaneous Generation, using some strange 'recipes'. This doctrine underpinned the activities of inquisitive people who, even in those surprisingly late days, 'created' mice and other nice little creatures such as maggots, lice and worms. Although Van Helmont was sophisticated enough to introduce the term 'gas' to chemistry, and was first to show that air was made up of a variety of different gases, he had what are to us very immature, in the sense that 'stork brings baby' stories of reproduction are immature, ideas with respect to these processes[2]: If a dirty shirt is stored with wheat for a few days, it 'transchangeth' into mice. In 1668, Francesco Redi[3] put a cap on this, quite literally, by placing lids on vessels – *"having well-closed and sealed them"* – and showing that this promptly stopped the generation of nasty creatures, such as *"..scorpions, flies, worms, and such like called imperfect by the scholars"*. He argued that the creatures had standard mothers. A conjectured vermin-creation meme became extinct and knowledge developed and became more sophisticated.

As an aside, consider for a moment a wider angled-view of how science has been changing – and maturing? – over the centuries. About a thousand years ago, all science was purely empirical, and it was concerned with simply, but meticulously, describing natural phenomena, such as the states of the cosmos – essentially the night sky at that time – as seen with unaided vision. Within the last few hundred years, a more theoretical branch sprouted, using mathematics to obtain models and generalization to provide, eg, foundations of the great engineering advances of industrial revolutions. The Law of the Pendulum gives a relatively simple example. In more recent decades, a computational branch of science has appeared, for simulating complex and often inaccessible phenomena, such as weather systems. Today, some would argue that we make heavy use of *data exploration* as a sub-activity within scientific exploration, and attempt thereby to unify theory, experiment, and simulation using data captured by instruments, or generated by simulator, which is processed by software, and large quantities of information and knowledge are stored in distributed computer systems. Analysis of databases and files, often being voluminous enough to be called 'big

SUPERINTELLIGENCE AND WORLD-VIEWS

data', is relied heavily upon by the modern scientist, using something called 'data mining' and possibly statistics. This new branch of knowledge discovery is still developing, albeit very rapidly.

Immature thinking does play an important role in the way we develop as individuals. For example, young children of school age are often introduced to 'nature study', which serves the very useful purpose of stimulating some, admittedly relatively low-level, investigation of naturally occurring phenomena. It opens the door, I believe, to much deeper thinking in later years. In my primary school days, for example, the keeping of nature study journals and the use of bird-watching handbooks was encouraged. This stimulated further exploration leading to the fuzzy border between nature study and science. Young minds could be exposed to the: *"...idea of science, method, order, principle, and system"*, set out by Newman, and the interfacing of this with knowledge from other sources, and from other disciplines. Also nature study was a welcome diversion away from rote memorization and recitation. To use a well-worn phrase, the study of the book of nature supplemented the second-hand knowledge from conventional books. It served to instil a mental discipline of a rudimentary sort and to: *"prevent a merely passive reception of images and ideas which in that case are likely to pass out of the mind as soon as they have entered it..."*.Tuition could, as Newman said:[5] *"...impress upon a boy's mind the idea ...of rule and exception, of richness and harmony"*. If the boy matured and got into the way *"...of starting from fixed points, of making his ground good as he goes, of distinguishing what he knows from what he does not know"*, Newman believed that he would: *"...be gradually initiated into the largest and truest philosophical views, and will feel nothing but impatience and disgust at the random theories and imposing sophistries and dashing paradoxes, which carry away half-formed and superficial intellects"*.

What about the *enigmatic* nature of knowledge? We can see it by looking briefly again at one of the modern Big Questions: viz, what is the origin of the physical universe, how did it get to its present state, and how might it end up? This falls within the domain of cosmology, which has advanced a lot in recent decades. Closely related to the Standard Model of Human Origins we used earlier, there is what I'll call a Standard Model of Cosmology, which is essentially a series of mathematical equations that are claimed to describe the universe as we perceive it today. For example, by a *sort of* reverse engineering, scientists try to 'model' how the universe started – with a big bang, its history since, and its future. In spite of the progress made, impressive as it is in terms of the density of squiggles produced, there are still many puzzles that remain about these matters. Alongside the mysteries of quantum fluctuations I mentioned in Chapter 8, another puzzle arises because the amount of matter that is observed within galaxies (called 'Visible Matter') does not explain the way the galaxies move. Some of that movement *can* be explained if we assume that galaxies contain extra 'Dark Matter' – matter that we can't see. But people still puzzle about this. Another enigma or riddle in this domain concerns the question of the universe's 'Dark Energy'. Galaxies are apparently being forced apart at an accelerating rate. According to the Standard Model, and observations, one estimate is that the total substance of the known universe contains 4.9% visible matter, 26.8% Dark Matter and 68.3% Dark Energy. That's a lot of darkness! Scientists puzzle about the distribution of dark energy and matter and how it interacts with and influences visible matter. Mathematically clever and creative 'models', intended to account for all things cosmological, appear frequently, but many of them have non-universal following among physicists in the field. Theorists have successfully come up with plausible answers to some earlier enigmas – for example, why the universe is not curved, why the matter in the universe is so uniformly distributed, and some other technicalities not covered by the standard Big Bang theory. This fits well with readings of cosmic background radiation and other observations. However, ever more theories and models keep appearing and the reason for this is that, despite the attractive features of current theories, many of the details addressed need assumptions that are not universally accepted. Other enigmas in physics arise famously in quantum mechanics, such as wave-particle duality.

There is also another definite sense in which knowledge is enigmatic at an even deeper level, causing our understanding to be impeded and darkened by being *in riddles*. As explained above knowledge acquisition, using methods such as induction, has been extraordinarily successful – for example in underpinning technological advance. Induction works! But can it be relied upon totally? With respect to the claims of human knowledge, as we've seen, *Postmodernist* thinkers and others have pointed out implications of variations in the contexts, orientations and even prejudices of explorers, and also to their intrinsic limits. Getting a handle on objective reality, might, they say, prove to be impossible, and we might have to be content with our fuzzy and otherwise somewhat unsatisfactory, FAPP-oriented knowledge, and completely give up ideas of finally verifying it against reality. Social, political and environmental systems, and even mathematics, can also have their own in-built riddles. The case of mathematical knowledge is particularly interesting, and salutary, given the existence of claims that it produces 'pure knowledge'. Philosophical differences have arisen in mathematics with regard to standards for proof, for example, in sub-disciplines, say, analysis, number theory or geometry. A controversy dating back to the start of the last century concerns a kind of mathematics, called *constructive mathematics*. It takes a different stance on existence to that taken in classical mathematics, where a proof of the existence of some mathematical object doesn't have to indicate a means of its construction or provide an example of it. A non-constructive proof in this case would prove the existence of the object without such demonstration. An example from economics is the existence of a 'Nash equilibrium' proved by John Nash in 1951, and there's an ancient one due to Euclid: *"There are infinitely many prime numbers"*. Other disagreements exist on relatively basic things, like the Platonic concept of two referred to just above.

We should not take this too far, though. For example, the current *zeitgeist* appears to support the rather curious phenomenon we saw in Chapter 6 – it is somehow seen to be attractive to deny, usually as a result of some kind of ostensibly rigorous enquiry, the possibility of the existence of objective truth. It is held that it is established beyond-doubt, as hard-fact, and really-and-truly true that 'truth' is subjective, that objective truth claims merely amount to ideological mumbo-jumbo. But, as we saw earlier, very little reflection is needed to see that this is patently self-contradictory. We would, by definition, not be able to use good, every day, honest reasoning as the way to establish that proposition – ie that there is no such thing as objective truth. Umberto Eco[4] vividly captures the annoyance at the brain hurt this evokes when he expresses his potentially violent reaction like this: *"What I continue to consider irrational is somebody's insistence that, for instance, Desire always wins out over the modus ponens;...but then...to confute my confutation, he tries to catch me in contradiction by using the modus ponens....I feel a desire to bash him one"*. While living with the unavoidable conclusion that knowledge dissemination is necessarily tainted by the fact that personal viewpoints might unconsciously produce 'spin', and that outputs of inductive and other forms of exploration are limited in various ways, my viewpoint here is that we *can* to some degree approach, or reach, the truth in some objective sense, although this does not mean that we can grasp it.

We also have to acknowledge the *partial* nature of knowledge, which is seen in the examples just now. Individual disciplines are limited in another sense – in scope as well as in extent. No individual discipline, such as one particular science, furnishes the whole truth. *Ipso facto*, not even the most knowledgeable polymath or guru could hope to have full knowledge of everything that has been discovered, even if all possible knowledge had been discovered already. And no-one claims that it has. It is becoming increasingly hard to keep up with even a specialised sub-discipline. Ask any academic. Going back to the early days of scientific exploration, natural knowledge was *universal* in important senses, but it is currently 'managed' in a sense by investigators who, unlike polymaths such as Newton, Bacon or even Da Vinci, are localized in their interests and expertise. Various threads are intimately connected, intertwined, and would have to be spliced together by an omniscient being if they were to form a single whole. This is why the need for the idea of a 'university' arises, and that very word itself is based on universal knowledge. To varying extents, it[5]: *"...teaches of plants, and earths, and creeping things, and beasts, and gases, about the crust of the earth and the changes of the atmosphere, about sun, moon, and stars, about man and his doings, about the history of the world, about sensation, memory, and the passions, about duty, about cause and effect, about all things imaginable"*.

Each individual's knowledge is typically quite narrow, even within a single discipline such as pure maths or physics. For example, Andrew Wiles, who transformed Fermat's centuries old Conjecture into a theorem may have contributed little to the theory of probability. Furthermore, the Gödel theorems could be used to show that no system of physics, for example, can ever be complete and remain sound simultaneously. So a global understanding eludes even the most outstanding individuals, and yet a synthesised view of the nature of things at some level is still sought. Being as complete as possible is important for a cosmos-view. If even one branch of knowledge is left out, the aggregate is seriously diminished. Another discipline, which needs its outputs, from outside that discipline's 'club', for example to invalidate apparent facts, would lose out. As Newman pointed out, *"...leaving out' disciplines would not be sophistication – it would be 'not knowledge, but ignorance"*. However Einstein and Picasso were geniuses, but worked in different disciplines. As Alexander Pope has written[6]: *"One science only will one genius fit; so vast is art, so narrow human wit."*

The intertwining of disciplines is very interesting, and it has deep roots in our predispositions in life. But how can we put our trust in methods that are so fragmented, which no-one understands fully? A lot of cross-disciplinary *trust* is needed, despite the fact that authority, and peer-review have their place in this. A kind of 'someone somewhere is keeping an eye on all of this' assumption is prevalent. It is a sort of folk-belief or default common wisdom that is entertained unacknowledged or even unconsciously by many, and this leads us nicely to our next consideration.

Faith. Like that inter-disciplinary trust, and over against, and at some distance from, the concepts under-pinning principled observation, induction and other knowledge establishment methods, there is the concept of *faith*. It is commonly associated with matters sacred. However, even in secular settings, there are people who, in practice, believe in the absolute necessity of faith in the sense I'm thinking of. Now the biblical term, *"the conviction of things not seen"*, often taken to refer to things 'not visually seen naturally by the agent', has been used to describe the concept of faith. But 'not seen' here does not imply or require physical blindness, but *lack of direct observation* through any sense. Children, along with scientists and others, believe in lots of things they have never seen with their own unaided eyes. For example some scientists take as an approximate fact, the existence of 10^{22} or so stars, and of the radio waves and magnetic fields we know surround us, but they have not seen many of them in the night sky. Even with a robust knowledge of mathematics, we might not 'see' how Andrew Wiles proved his famous result mentioned just above. Yet an evaluation of the evidence at our disposal and assimilable by us, and the testimony of authorities, has led us to believe in these things. That's faith in the sense we're using the word here. Some would say that this is anti-intellectual – that reason stands on one side opposed to faith on the other. But for faith in the sense I'm using here to exist at all, the mind, emotions, reasoning and all, must be involved. It entails both the perceiving and the understanding faculties. In this sense, it is simply entirely wrong to think that faith and reason are in opposition. Faith should, of course, be reasonable, and it should be possible to give a reasonable justification of why we retain it, and so in the sense here, faith is not vague or blind or irrational. It is nothing to be ashamed of and brains are not 'checked in at the door'.

Let's look now a little more at problems raised as we seek to *live with indeterminacy and uncertainty*. Even non-scientists will have heard of Heisenberg's Uncertainty Principle. Heisenberg produced his famous formulation in 1927 as the result of a thought experiment, which entailed trying to measure the position of an electron using a very powerful microscope. He established that it is impossible for an experimenter to know both the position and the momentum precisely. In fact it is impossible to measure properties of the system without disturbing the state of the system. The energy of the light in it, for example is *quantized* – its size is restricted to only certain discrete values – and this sets a limit to how small the effect of the disturbance can be. So the very act of measurement itself introduces unavoidable uncertainty in the measurement. The implication of this is that we cannot know the present state of the world in full detail. This, alongside other insights from Quantum Theory and fields such as Chaos Theory and Gödel's Theorem, in turn means that we can't predict the future with the absolute precision we'd perhaps like.

Heisenberg's principle predated any experimental equipment or approaches that could test it at the quantum level. The idea has only been put to the test in the lab recently, and there are claims that 'weak measurements' show that the principle was never quite right. There is a still a fundamental limit of knowability of all details of physical systems, though. To find out with certainty is still seen as an unattainable goal. However, we don't despair, as knowing without certainty and believing inconsistent things from time to time, is something humans have done throughout history. Some say it is the mark of some of the higher capabilities we've talked about to cope well with this[7]. *"The test of a first-rate intelligence is the ability to hold two opposing ideas in mind at the same time...and still retain the ability to function"*.

Living with Inconsistency. Some researchers see learning as partly being a process of replacing or adapting conceptions that are mistaken or are out-dated with more appropriate new knowledge – re-interpretation of what the learner already knows in the light of happenings, situations, events, activities, and other phenomena. Maybe the environment has changed, or the learner simply gets some corrected data. New information – eg from material presented in an educational setting – means contradictions of varying longevity often present themselves. Changing world-views and perceptions that are localised in some ways, perhaps of the Relativist kind met in Chapter 6, for example, like 'drinking alcohol at home is evil', can therefore lead to inconsistency. We, and other researchers in this area, have argued that it should be acceptable for learners to accommodate some contradiction.

This is illustrated nicely by young humans, who may be quite comfortable as they retain some 'working contradictions'. An

SUPERINTELLIGENCE AND WORLD-VIEWS

obvious source of these can be (immature) misconceptions, which they believe strongly. For example, in a study reported by William Philips[8], many fundamental facts like the shape of the Earth were thought to be known to the young subjects, but in fact they were equivocal. Younger children can accommodate contradictory beliefs about things like Santa Claus and fairies. They are quite good at make-believe. This has the advantage of allowing the children to recognise that things that they do not yet understand may appear or occur. They are often happy to foster particular beliefs that are inconsistent with the rest of their knowledge, and they can act as if some strange things were true under specific circumstances.

They are in good company in employing this cognitive device; philosophers and scientists among others sometimes treat contradictions in ways that are somewhat similar. For example, *dialetheists* take the view that some contradictions are true, or that something can be both true and false at the same time. Consider the infamous Liar Paradox example: If we make a statement, say called Statement 1, as follows: *Statement 1 is false*, we have a clear contradiction – Statement 1 is both true and false – a case of bad self-reference, not unlike the one we saw with some instances of relativism earlier. This paradox and others indicate that common sense seems to give us licence for holding contradictory beliefs.

Scientific theories often turn out to be inconsistent and in everyday life people contradict each other, and even themselves – frequently. Interestingly, this does not mean they believe just anything, as a standard rule in logic stating: 'a contradictory statement implies an arbitrary statement' would lead us to believe. The standard rule, rarely used in practical situations, is captured in the quotable Latin phrase, *ex contradictione quodlibet*. We'll not delve too far into this tricky subject here, but just mention a few developments in recent times. To get around the problems the above rule presents, as well as some inconsistencies that can occur between items in a knowledge repository, 'para-consistent logics' have increasingly become available. These tolerate contradictions but do not permit inferences of any old thing from inconsistent premises. Often rhetoric using such things as plays on words and metaphors is being used, or the definitions of terms are sometimes deliberately left vague. So we can say things like: 'you can't dip your toe into the same river twice', and yet you *can* go to a river and do it – the meaning of 'same' here is what causes the problem. Of course, intuition still tells us that only in exceptional circumstances should contradictions be accepted. *Ceteris paribus*, a derived contradiction still generally a reason against a belief.

Humans are natural explorers. They begin carrying out 'environmental study' that's richer than ESA's from the moment of birth, largely based on inductive methods. They seek patterns in their observations, and pre-school come up with many early empirical insights or proto conjectures. They might establish 'laws' FAPP in the world in which they find themselves such as 'my baby-cup bounces on the floor without breaking', and they might over-generalise this to say, for a while at least, 'all cups are unbreakable when they fall'. These immature, naïve understandings often provide shaky platforms for developing more sophisticated understandings but they can interfere with subsequent learning.

For early SIs' mechanical learning systems to be useful in real world situations, as well as coping with partial, enigmatic and immature knowledge as we have to, they would need features for dealing with complexity and change. As well as the obvious need for efficiency of operation there would be a need for continuous adaptation in changing circumstances if the agents were not to be simply playback. And it would be necessary for humans to understand any novel outputs and models that the system comes up with, for using and checking. Hopes of reaching the levels of efficiency required are pinned largely on numeric and statistical methods of learning as traditionally studied in the Machine Learning literature.

Our human ways of acquiring and changing concepts are not yet well-understood, and automated learning mechanisms that could help explain them are actively sought. Carey[9] proposes employing existing non-domain specific concepts for this. These give us widely applicable conceptual frameworks – our 'intellectual road maps', presumably including causal networks and other classification devices that are said to be present very early in the lives of us humans. They may include aspects of the innate 'core domains' of intuitive understanding of certain concepts such as those in naïve physics, naïve psychology, naïve biology, naïve number structures, and naïve spatial concepts that we saw in Chapter 11. To take an example, Carey[9] suggests that in bootstrapping the concept of 'living thing', initially young children start with an undifferentiated notion, maybe combining 'alive', 'existent' and 'active' in a single class. When they start to learn more about pets, people, germs and plants they readily pick up simple causal facts such as 'my puppy, my carrot plant and I are all alike in that we all need food and water to keep alive and grow', and this eventually gives a grasp of the concept

SUPERINTELLIGENCE AND WORLD-VIEWS

'alive', uniting animals, plants and humans, distinct from those other concepts, 'existent' and 'active'.

Technologists who seek to design exploring robots, perhaps aspiring to eventually reach genius level, can learn from such research into human learning and behaviour. The learning agent must realise that the concepts they hold may be no longer applicable – triggering observation and exploration (see our exploring robot experiments described in Chapters 5 and 10). The learner must grasp the interplay between experience and the new concept to a degree that opens up unforeseen possibilities of gaining some benefits for the learner. The new concept must have potential be 'interesting' or 'useful', and (ultimately) be consistent with one's world-views, one's other knowledge and one's past experience.

Handling Inconsistency and Uncertainty. At a more collective level than individuals growing up, there are many historical, scientific and religious examples of humans believing inconsistent things. These can be dealt with using concepts of inconsistency handling. We can see parallels between this and other common and accepted, but hopefully only temporary, contradictions, and perhaps we could tackle some of those by allocating *belief strengths* to the alternatives, such as: *[Santa and not Santa]*, *[falling cups don't break and falling cups break]*, *[the earth is flat and the earth is not flat]*, and *[Statement X is true and Statement X is false]*. The contenders in each case could be allocated a 'portion of available belief strength' in accordance with the believer's respective confidence in them. The same applies to cases where there are more than two exclusive and exhaustive contenders to believe. Modelling uncertainty by allocating 'confidence levels' like this is good for more than intellectual satisfaction – it is helpful in making decisions FAPP.

Non-monotonic reasoning allows agents to make inferences that can be undone. Tentative conclusions *can be retracted* in the light of new information. Uncertainty can also occur in general quantified statements like one we saw earlier in this chapter: 'all swans are white'. Reasoning has to deal with uncertainty in both individual entities, such as tuples in a database, like a patient in a medical case, and generic information such as rules. For example, 'if the patient's temperature is high, then the patient has flu'. Of course there just might be (many) exceptions to that particular rule, due to there being other things that can cause high temperature, and to the general variability of circumstances, and all of this introduces (quite a bit of) uncertainty. Moreover, different information might be available from different sources, such as expert opinions of doctors in the medical case, or imprecision in, or simple lack of, facts. Formally, probability theory facilitates well-accepted and quite sophisticated reasoning based on the allocation of uncertainty levels.

One generalisation of probability theory that is often used in AI is called Dempster-Shafer theory or Evidence Theory[10]. A *Belief measure*, or 'mass', between 0 and 1, is allocated to a particular subset say of the contenders say in the medical case above, when evidence is available, perhaps from a lab test or expert opinion. There is a special rule called the *orthogonal sum* that allows combination of masses of evidence obtained from independent sources on (subsets of) the contenders,[10].

It has been shown many times that this sort of calculation is powerful and has advantages over the standard probability methods for some more complex applications. In my own group, for example, over the years we've used it widely in our research – on aircraft wing material choice, writer identification in noisy and obscure documents such as Bach manuscripts, and even trying to predict the final scores of the 31 matches in a football tournament – the 1996 European Football Championship. Its potential usefulness in more mundane AI applications is undoubted, but it comes nowhere near giving a model of how we humans do our uncertain reasoning. So we don't yet have a totally tight and final way for SIs to handle uncertainty in reasoning post-Singularity.

References

1. Deutsch, D., *The Beginning of Infinity*, Penguin Books, 2011.
2. Van Helmont, J. B., *Ortus Medicinae*, 1648. quoted in http://rationalwiki.org/wiki/Abiogenesis#cite_ref-3
3. Redi, F., *Experiments on the generation of insects*, The Open Court publishing company, 1909.
4. Eco, U., On the Crisis of the Crisis of Reason, in *Faith in Fakes*, Vantage, 1998.
5. Newman, J., *On the Scope and Nature of University Education*, Everyman's Library, 1915.
6. Pope, A., *An Essay on Criticism'*, Part 1, Line 60, 1711.http://www.poetryfoundation.org/learning/essay/237826?page=2
7. Fitzgerald, F. S., *The Crack-Up*, (1945). New Directions, originally Esquire magazine, February 1936.
8. Philips, W., Earth Science Misconceptions. *The Science Teacher*, 58(2), 1991.

9. Carey, S., Conceptual Differences Between Children and Adults, *Mind & Language*, Vol 3 No. 3, 1988.
10. Guan, J. W., Bell, D.A., *Evidence Theory and its Applications, Vols 1 and 2*, Springer, 1991–92.

Chapter 13

Towards Apocalyptic AI – How far have we still to go?

Prophecies predicting the end of the known world have persisted throughout most of the history of civilisation. The idea of an Apocalypse has certainly been around in religious circles from at least the time of the biblical prophet Daniel when Belshazzar reigned in Babylon two and a half millennia ago. Often the more ominous aspects of this prediction – mainly the catastrophic retributive destruction of the world, which is rather more negative than the OPT picture – are what people think of when the adjective 'apocalyptic' is used, but there is a nice side to it as well. Indeed the word *apocalypse* comes from a Greek word roughly meaning 'an unveiling'. Apocalypticism as a belief is prominent in contemporary theology, particularly among Christian fundamentalist groups, who look forward to an 'end of the world' scenario involving an 'eschatological kingdom', often taken to be in a heaven over which God rules, and in which he dispenses blessings and finally works in harmony with humankind. The vision of 'The End' in Chapters 6, 7, and 8 centres around one in which a future of ubiquitous computation that has somehow acquired significance and meaning leaves a lot of 'ifs and ands'. On that view, just as in that of Daniel just mentioned, and just as in many other somewhat similarly positive visions, two big widely-recognised fascinations and pre-occupations of reflective humans, viz, desired immortality of the individual and the need for a purpose in life, are seen to be satisfiable. However, the prospects for this coming from or being contributed to significantly by any projected computational developments, and the chances of those happening any time soon, are not as rosy as many claim. Although cyber-sophistication is increasing dramatically all the time,

SUPERINTELLIGENCE AND WORLD-VIEWS

and it has expanded to engulf many people's worlds, there is still some way to go if it is to have the features that will allow it to determine the future limits and potentials of our cosmos as a whole.

The Singularity. AI based capabilities can be found these days in Internet search engines, autonomous vehicles, game-playing, quiz-playing systems and image processing for security applications, to take just a few examples. But as Rodney Brooks, formerly head of MIT's Humanoid Robotics Group, put it neatly, *"...you can't engage in an interesting heart to power-source talk with any of them"*. Yet, that is...! As a result of the progress made to date, optimism among self-styled latter day counterparts to, say, biblical prophets, though tending to be somewhat more secular, has increased to the point where it has buoyed up a movement called *Transhumanism*, whose followers anticipate desirable, deep-rooted, engineered changes in our human physical and mental make-up. This is to be distinguished from *Posthumanism* which postulates the advent of SIs. Transhumanists predict ultimate life extension by uploading our minds and consciousness into computers as Tipler and Deutsch suggest, and extensive life enhancement by linking flesh and metal in increasingly clever ways. Brooks has pointed out that this is already happening, and neuroprosthetic technology is said to increasingly link the 'meat' brain with silicon devices such as robotic arms and computer cursors[1]. And some prophets say that virtual reality will gradually become indistinguishable from real reality, and it will eventually interact more directly with our human nervous systems. And then there is Whole Brain Emulation and so on that we saw something of in Chapter 2.

Humans and Transhumans may, it's claimed, stay in control of any SIs. In 2008 Brooks[2] wrote that human-machine composites will *"...always be a step ahead of them, the machine-machines, because we will adopt the new technologies used to build those machines right into our own heads and bodies"*. Transhumans, and humans might therefore be able to impose our human world-views on SIs or artilects[3]. The Apocalyptic twist on this is that some see extreme Transhumanism as a way of conquering death. As we saw in Chapters 6, 7 and 8, Tipler is one of those dreaming of Transhumanism as a path to the desired immortality of the individual human and the best dreams include scenarios that satisfy the need for a purpose in human life.

For Posthumanism there are different takes on likely details. Will there be continued gradual slow advances, or a distinguished, crisp, catastrophic 'great leap forward' change as the result of progress in artificial intelligence, abruptly ushering in a new world? The received wisdom seems to be that such a leap, if possible at all, could come about in the next 100 years or so, but some specialists say it is only about 20 years away. Hans Moravec, who was one of the first to write about this in a relatively realistic way, once estimated 2018 as the date. Significant software developments will be necessary for this, but those who accept the evolutionary stance of, say, Teilhard, for example, say confidently that we can emulate any evolutionary leaps in raw nature in due course, given the will and resources. So the claim is that, in the search for cosmic purpose, an important part of a cosmos-view, Tipler's sort of visions and conjectures concerning machines, resources and human beings are attractive. The technical/scientific feasibility of this can appear to be sketchy and it can be seriously questioned, however, and it raises interesting philosophical and theological problems, as indicated earlier. Tipler seems to have done his checking carefully, but there always remains nagging doubts about stretched concepts that are apparently 'flaky' and are certainly inaccessible to most of us. Having said this, at a deeper level these visions might well, and to some minds already do, represent a competitor, or at least a challenge, to other forms of human belief and faith, especially those with a more traditional apocalyptic character. Some optimistic Transhumanists look forward to a time when our re-embodied minds will persist and learn more and more about things in general – 'life, the universe, and everything'. Or perhaps all this will take place virtually.

Of most interest to us in this book, of course, is the SI scenario, and SIs' likely post-Singularity relationships with us 'more meat than silicon' agents. The possibility looms large that we might have to pass the baton on to a 'new species', the 'pure machines'. Just suppose for a minute that the resulting entities as they emerge refuse to comply with the wishes of their originators, and maybe, like W*22, develop relationships with standard humans that are not those that optimistic apocalypsists, for example, hope for. They may not like what they know or find out about us – and vice-versa, of course. These uncertainties give grounds for some concern about what computing researchers are doing and how mankind can keep in control of the developments. This all impacts on the interests of the great world faiths, of course, and this would probably lead them to decry Apocalyptic AI goals.

Another very big question to accompany those raised by all of this is: What advantages do we have over artificial intelligent agents now, and could we be assured of, or work towards, maintaining that

SUPERINTELLIGENCE AND WORLD-VIEWS

advantage? So: do we want to put a stop to all of this development? Or perhaps we can make efforts to ensure that the new developments keep humans in the driving seat?

One possibility for keeping humans in control, is to jealously retain our 'humans-only' features – such as emotions, ethics, creativity, consciousness and even a sense of humour. The SIs might, of course, acquire them, perhaps accidentally or by some more sinister means. And are humans going to decide, for example, what the *values* of SIs are going to be, as Brooks seems to hope? Confining them to things like humility, empathy, self-control, fairness and duty, while masking out any tendencies towards things like selfishness, excessive 'drivenness', and arrogance might be the way forward. The aspiration is that well-intentioned individuals of all sorts can agree to disagree where necessary. The utopian take on this is that individuals can be tolerant and level-headed about the choices that others, artificial and natural, make. However, anyone looking at mankind's history to date might just say that this is sort of thinking is just a little bit wishful! The only option might be to hope for the best.

In the real world, even if humans were to decide, as a species, that we still want to gradually mechanise ourselves, or if endowing computers with our minds or some other sort of 'human essence', or even producing ('better than us') SIs, is attractive, the very important fact must be faced that *there are still a lot of problems to solve*, despite the fact that people have already been working energetically on these problems. Recently Google has become increasingly interested in this sort of thing, reportedly buying up several robotics companies like DeepMind, as we saw in Chapter 2. To get anywhere near SI, though, those researchers or others will first have many very hard issues to address before we get anywhere near the things such as emotions and creativity in machines.

I'm not focussing on it here, but one big technological challenge in building artificial agents with superintelligence and world-views is that of developing systems of sufficient scale and power. Recent advances in computational power, often expressed in this context as high speed, can encourage SI enthusiasts, but one downside is that supercomputers use huge amounts of electrical power. A small town could be serviced by the power consumed by the former world's fastest super-computer, Tianhe-2. Tianhe offers 33.86 petaflops, ie $33.86*10^{15}$ floating-point operations/sec (FLOPS). Its energy usage would need a small power station to generate. And those megawatts of electricity produce heat, which needs more power to cool it down, and cooling systems become important.

Incidentally, when people have studied what we'd need to emulate all human history, in a *sort of* virtual reality world, somewhat as Deutsch's suggestion that we saw in Chapters 6, 7 and 8, various estimates have been obtained. Let's use one well-known estimate by Oxford University philosopher Nick Bostrum for illustration, namely around 10^{36} operations in total. So Tianhe won't do. However, a rough approximation of the computational power of a planetary-mass computer is 10^{42} FLOPS, so, on these figures, a single computer of that power could accommodate the simulating of the mental history of humankind many times over, at least conceivably. Tipler's models and calculations come into the picture again, of course, but remember, as I've said before, we can't even control the weather here on Earth, so we seem to be a long way from doing this.

Anyhow, hardware and exhaustive search alone are nowhere near enough for SIs, as we saw briefly when looking at games-playing software like AlphaGo, Watson and Deep Blue. Flexibility, functionality and cleverness are needed in the digital cyberspace world to facilitate the proposed 'higher than BAA', genius level agents. It is tempting but very misleading to claim that any problem can be solved by throwing enough computational muscle power at it. Superintelligence requires good software. Here we simply point out that there are architectural and other issues to be addressed. Some systems have been spoon-fed facts, rules of thumb and heuristics, over a long time by a team of humans. Watson's paradigm involves less tailored human input, making more use of learning and text understanding systems and existing repositories – more bottom-up/work-forward. However as seen earlier, a very smart *driving superstructure* as presented in Chapter 2 might be a pre-requisite to pushing ahead with superintelligent innovation. New modules might be added for domains to parallel chess, maths, etc, but the design of an appropriate superstructure is the key to achieving SI.

It is certainly 'reachable' that, given our relatively limited perceptual equipment, humans won't be able to keep up for very long with AI developments. Clearly important, and hard, ethical and other value-related, questions, for example, can be seen to be raised, so a multi-disciplinary approach to SI design is called for. Moreover, the projected functionality will be based, at least initially, on devices, rules and procedures set up by the machines' human designers and implementers, and it should be remembered that the main goal of the vast majority of serious computing researchers working on these sorts of problems is to make information-handling and problem-solving tools for mankind – not to achieve the Singularity. Furthermore, we will list below a set of definite lacks in the functionality of current offerings of computing machines that need to be addressed if the Singularity is to materialise. Perhaps steps could be taken to stop some of these gaps being filled.

Our focus in the rest of this chapter is to look at some such functional capabilities and characteristics, arguably needed for this sketched-out future, that are conspicuously missing from most machines currently available or envisaged in the near future in this domain. Consider the situation in *Ex Machina*, where Ava simulates human *emotions* from knowledge of humanity gleaned from cyber-information using a fictional search engine, called Bluebook, and learning from this alongside some limited observation of humans. There's not much emotion shown when, say, she/it is physically tussling with one of the humans, Nathan. Furthermore, as well as, say, a spiritual dimension and the values mentioned above, advanced machines of the kinds we've been thinking about will need consciousness. There are still a lot of questions to be answered here, but they have had a lot of coverage elsewhere[4,5]. How do spiking neurons and other physiological activities give rise to conscious experiences inside humans? How does the brain: "*...thinking meat... loving meat...dreaming meat*", to use some of Terry Bisson's words[6], put the heterogeneous inputs from my various physical (meat) sense organs and unconscious visions together, in order to make up a single subjective experience? To obtain any facts about the nature of consciousness, or about an individual agent's mind, or about minds in general, the subject concerned has to be asked questions. Then what about some other things that Ava does not demonstrate much of in the movie, such as *creativity*? A *sense of humour* might also be needed for a minimally satisfactory 'heart to power source' conversation. I've touched on some of these things in earlier chapters but there are a few specific aspects of each I'd like to cover. So let's look briefly at where we are with each of these in turn.

Emotion. First of all, let's consider emotion, and note that this attribute is at best minimally present in any current AI, but it is considered to be very significant for humans. We can express emotion, and react to it in others, but computers can't. A 'Turing test for emotions' would be hard to devise because of the difficulty of saying when a clever robot is pretending to emote. Maybe just making them sensitive to our emotions is all we need? They would then have human interests 'at heart', rather than be simply indifferent to any emotions that do not have a 'pay-day' for them. Anyhow there would have to be a wide spectrum of emotional requirements to match the wide range of applications we can envisage. Some of the attributes might not be strictly needed for car-assembly robot, but for a child-minding mobile agent, for example, they would be essentials.

Consider a servant or companion mechanical agent. Do we really need a computer that pulls its artificial hair out if it doesn't get its own way? Of course, we don't necessarily need our machines to, say, leave us to go to work for that 'gap year' for some charity in the third world, nor will they any time soon. And if we do choose to allow robots to be given or somehow acquire some basic emotions of humans, we'll have to guide them. Complex emotions like being envious of someone, where the emoter has to recognize that that 'someone', perhaps a competitor, has something they want and change attitude accordingly will be hard to produce and control.

Emotions are important in human decision-making as well as in basic interaction, and it's widely acknowledged these days that our decision-making isn't purely logical[7]. The fact that it's also emotional has been established by findings in neuroscience and other disciplines. For example, a prominent team researching in human neuroscience, led by Antonio Damasio, made a seminal discovery in the early 1990s about the neural correlates of emotions. Impairment to a particular brain region, the ventromedial prefrontal cortex (vmPFC) can affect social behaviour profoundly[8,9]. Previously well adapted individuals may display a newly acquired difficulty in deciding on relatively simple matters pertaining to their lives, or they may be unable to follow social conventions. Remarkably, the patients' basic intellectual abilities are generally unaffected, in that they use language normally, and their learning and reasoning and attention capabilities remain sound. Moreover, they can, in fact, consciously make moral judgments on hypothetical situations presented to them 'at arm's length' – as in thought experiments. But when it comes to similar situations that they are involved in in their own lives, they: *"exhibit an abnormally high rate of utilitarian judgements"*. When shown emotionally charged pictures they show reduced: *"empathy, embarrassment and guilt"*.

In fact, at the sharp edge of decision-making, even with what we believe are crisp logical decisions, the ultimate tipping-point of choice is arguably, and sometimes evidently, based on emotion. Damasio's work seems to indicate that individuals making judgments and decisions involve emotional qualities in a primary way, and that all such

SUPERINTELLIGENCE AND WORLD-VIEWS

cognitive operations and expressions are in fact affected by emotion. Values such as empathy and sympathy play crucial roles in much of our human social interaction. So feelings would be very important for W*25 and successors. However, they won't let emotions get in the way of making smart decisions, unlike some people. Even people with intact vmPFC operation sometimes have serious negative interference from emotions. For instance, someone has a row with his wife, gets to work and the residue of emotions surviving from the row influences what he does in his office. Incidentally, many of the characteristics being considered in this chapter – the 'lacks' in modern 'smart machines' that we're considering – are interrelated and operationally intertwined. For example, emotion and consciousness are linked, and damage to the vmPFC has also been connected with deficits in detecting irony, sarcasm, and deception, which are not unrelated to having a sense of humour.

Alongside our 'how much emotion reading?' question for machines, there is an open question for humans: How can humans ensure that they are proportionately swayed by the emotions being experienced, and that unwanted interference from those emotions that aren't related to the decision at hand? 'Emotionally intelligent' people remove emotions that have nothing to do with the decision, but they make sure that they keep other emotions – viz those needed positively for the reasons stated above. They are protected from the worst tendencies to permit overwhelming emotional swamping. Current computers don't have emotions intruding in their decision-making, but the answer to this question is of value for SI design – for helping SIs to work in empathy and harmony with humans. That important aspect of emotion-handling between man and machine needs to be improved, but the answer to our 'how much emotion?' question is not clear yet, and the provision of necessary and sufficient machine emotions still seems to be some way off.

We should be able to control our machines better during our interactions with them than we can control our friends, colleagues and people in general, and probably better than we can control ourselves. However, even getting machines that are purely playback-emotional, with appropriate attitudes and behaviour towards humans, the environment and other machines is very hard in practice. Efficient, effective algorithms and systems that, where appropriate, govern flexible and adjustable mechanisms to take account of or include emotions, will be hard to design.

Letting machines with under-developed emotions loose to change aspects of the world could be dangerous without some restraint.

We're talking about a 'letting loose' that exceeds even that enjoyed by the autopilot of, say, an Airbus A320. 'Autopilots' use largely playback systems, and they have to ensure, for example, that in operation the aeroplane has to (usually!) stay inside certain limits – maybe prohibiting actions which would cause a stall. Every movement is computed and calculated, in reactive, but still definitely very tight 'playback', mode, by the autopilot, so they're not 'let loose' in our sense. Another example of some freedom of machines is in current ground-controlled semiautonomous unmanned assault vehicles, such as General Atomics Aeronautical Systems (GA-ASI) MQ-9 Reaper used by the US Air Force, and the earlier MQ-1 Predator. But they are also far away from the sort of 'letting loose' we're talking about.

If rampant, self-perpetuating machines are let loose, the horrified cry might go up, 'Who let the 'bots out?' The dreaded technical explosion that runs out of control and leaves humankind in the dust can be avoided. But will it be? Humans have to decide whether or not they prefer creations like these to make emotionally-sterile decisions, or, at the other extreme, whether they will allow their decisions to be made under the influence of, possibly immature, emotions. If it turned out that the progress towards superintelligence couldn't be stopped, how could the technologists and philosophers ensure that they include appropriate emotions in their designs? The assumption has to be that they still won't voluntarily choose, as presumably the early W*s, those generations before the ones in our little 'play' in Chapter 1, did initially, to go along human-specified paths on a final journey – from the point where they are made obsolete by the arrival of the next generation to their final resting places in the artilect cemetery.

Creativity. Another important manifestation of free agents which all current 'intelligent systems' lack is *creativity*. Creative thinking is arguably the highest mental faculty seen in humans. At its best it allows the thinker to conceive of futures that have never ever been experienced. Einstein provided insights into the workings of nature and Picasso produced strange shapes with strange colours in paintings on topics that included war, an old guitarist and some girls from Avignon. Creativity has had enormous transforming power in all aspects of civilization, and yet it is ill-understood and discussions of its nature can be confusing and feature some highly imaginative and perhaps incautious inputs. If one tries to see how it can be acquired by W*21, there is not a lot of detail to work with in terms of sound algorithmic or other mechanisms possessed by and manifested in the work of certain humans, as we saw in Chapters 4 and 5.

SUPERINTELLIGENCE AND WORLD-VIEWS

We looked at Emily's 'creativity' earlier, but I want to look, very briefly, here at what the prospects are for the design of robots and other artefacts that aspire to this more generally. In humans this is closely related to *imagination* – the generation, evaluation and handling of conscious mental representations of entities, behaviour and consequences, which are at least, very distant spatially from any sensory origin. Igor Aleksander[10] has said humans "...*live by, continually add to, and trade our imaginings*". Imagination is needed to visualise to some degree of accuracy or feasibility, alternative futures or some other aspects of a cosmos or world never experienced. SIs would also presumably need this if they are to be *spiritual* in all the ways that humans are, and it would include dreams, thought experiments, desires and vision, along with associated emotions, for example.

How could a system that had all of mankind's knowledge that's currently stored anywhere on computers, and available and accessible more or less instantaneously, push back the frontiers of knowledge in the way that human researchers, for example, do? Emotions would of course be invoked if they were to ask questions that are equivalent to, say, 'what do I feel about this music or that painting?' The system could have access to the knowledge bases of Watson, Emily, and others, and it would be able to access the Internet and other sources of information and knowledge. This again suggests the use of our superstructure of Fig 2.1 above various specialist systems. However, to emulate the sort of inborn drives that humans typically have that's self-triggered to visualise or simply 'reach' increases the challenge. Subsequently deciding on desired achievements and competences, and impelling them towards achievements and competences, would also be hard to arrange. Could Cepheus, as it now stands, have the idea of trying to work out ways of playing some other game such as Bridge? Geniuses, as we saw in for example Chapter 5, have a special gift, probably suitably encouraged and honed with experience, for dreaming up good ideas. That starts by *asking the right questions*. I've supervised many doctoral students – nearly 40 – and they showed considerable variation in this particular respect. All were very well prepared, assuming that they would not be accepted as graduate students if they did not have sufficient qualifications in the right area. But in many cases the abilities to combine, rearrange, and distil insights could not be predicted from their educational qualifications. It's one thing having a vast amount of knowledge that you can get down on scripts in exams, or find in 'big data' repositories, but quite another to fully understand a subject and from time to time say, 'Aha! Insight!'

There has been productive work on automated scientific discovery, knowledge discovery from databases, number theory, and other mathematical domains such as geometry[11, 12]. Programs like AM, and Cyrano[13] are well-known examples from mathematics which have produced some interesting results, but, while impressive in their own way, they are somewhat limited in scope, and heavily search oriented, compared to the sort of creativity we're looking at. For example, AM and Cyrano make conjectures but neither of these programs considers empirical data. There is other research that seeks to provide ways of finding, from known concepts or simple heuristics, mathematical objects of interest and proving theorems about them. There are also programs that generate examples, mathematical databases, and also computer algebra systems. Alan Turing once wrote[14] that: *"all intelligence is search"*, and that is fine if you know what you're searching for, how to recognise what is of interest and where to look. Clever methods of focussing and speeding up the search will not *per se* lead inexorably to superintelligence. And for machines to really innovate in various domains, the question of more widespread creativity remains to be addressed.

Are there other things we can try to copy from what we've seen of how exceptional researchers, inventors and other creative people innovate in Chapters 4 and 5, for example? There is, of course, a possibility that by some random fluctuation in their design the computers will come up with ideas that might be alien to us. But a more satisfactory first step could be to extract or replicate patterns of questions and innovative tricks or heuristics used by human geniuses – genius-tricks – or even those used by somewhat lesser mortals, when innovating. Really smart humans like Einstein and Newton, Aristotle and Plato, generate useful queries and problems over and above given objectives and goals, and pre-dispositions. The question they all have to answer is: 'What problem will I solve next?' As we've seen, often this is not done in a very systematic way – Koestler has called this sort of progress in the specific field of cosmology, *sleep-walking*. And this pattern is not an isolated, rare occurrence in scientific discovery. The Irish physicist John Stewart Bell expands on this before applying it in his own field[15], saying that Koestler recognised the contributions of people like Copernicus and Galilei. But saw them as being: *"... motivated by irrational prejudice, obstinately adhered to, making mistakes which they did not discover, which somehow cancelled at important points, and unable to recognize what was important in their results among the mass of details"*. Koestler's conclusion was that: *"...they were not really aware of what they were doing...sleepwalkers"*.

SUPERINTELLIGENCE AND WORLD-VIEWS

The geniuses listed above and others also added quite a bit of, presumably highly selected, and somewhat catalytic, 'something' to formal reasoning and knowledge systems. Perhaps well-formed formulas, axioms and rules, etc, could be supplemented by the addition of *some* informal rules of thumb, conjectures, and even some highly disputed or dubious matter?

There are some things that can be considered in addition to the 'search' methods mentioned just now, as a matter of course, and maybe even formally, in aspiring to innovation in machines. One example of shaking things up that humans use is *bi-sociation* where parts of other problems are combined, or solutions for analogous sub-problems are re-used. An 'analogue management' system or an 'overlap detection and adjustment' sub-system would be useful as part of a prototype artilect. Many years ago, for some problems involving optimisation in computing research I, like many others, used 'genetic adaptation', paralleling operations from genetics, such as invert (or otherwise rearrange), randomise and cross-over. And I've used 'simulated annealing', paralleling the physics process of producing crystals from a melt for the same sort of problem. Another 'shaking-up' method is to use *construction* – adding something into the picture that helps to make new insights fall out. And we can try to look at things in new ways, expand or contract, generalise or particularise – perhaps to extremes. We can say things like: 'Suppose the causal relationships I've assumed do not exist or there are some that I've ruled out too soon'; or 'something I've thought to be one kind of thing is, in fact, not of that kind'. Or 'what if I've mistaken the interrelationships between components of a system?' What's actually being done here is contributing to a *sort of* informal hypothesis generation, and it could be considered for inclusion in computerised systems.

The French mathematician, Henri Poincaré, wrote about trying to fathom the mysteries of certain human creative processes[16]: *"The genesis of mathematical creation is a problem which should intensely interest the psychologist. It is the activity in which the human mind seems to take least from the outside world"*. This is reminiscent of some of Einstein's comments in Chapter 5, for example. Poincaré goes on to give some details of a particular mathematical discovery process in his own experience that he found quite mysterious: *"For fifteen days I strove to prove that there could not be any functions like those I have since called Fuchsian functions. every day I seated myself at my work table, stayed an hour or two, tried a great number of combinations and reached no results"*. One night when sleep eluded him, *"Ideas rose in crowds; I felt them collide until pairs interlocked, so to speak, making a stable combination"*. By morning mathematical thought allowed him to reach an important stage in the discovery process, but, interestingly, he was engaged in non-mathematical activity when he got the next step forward: *"... we entered an omnibus to go some place or other. At the moment when I put my foot on the step the idea came to me, without anything in my former thoughts seeming to have paved the way for it..."* He did not verify the idea: *"that the transformations I had used to define the Fuchsian functions were identical with those of non-Euclidean geometry"*, there and then, but; took his seat and resumed his conversation. We still have some way to got to understand such processes and those of, say, Einstein and Picasso, to the degree needed if we are to add them to our SI's superstructure.

More recently another mathematician, Cédric Villani, who, incidentally, was appointed as director of Institut Henri Poincaré in Paris in 2009, also described the sort of experience he had when making an advance that led to him being awarded one of the highest honours there is for mathematicians. In his book[17] he wrote about a morning in April, 2009: *"Uhhhh! Man is it hard to wake up... Finally with the greatest of difficulty I manage to sit up in bed. Huh? I hear a voice in my head – You've got to bring over the second term from the other side, take the Fourier Transform and invert in L^2"*. He also found the key 'aha' aspects of the process mysterious, and commented on the lack of mention of these 'aspects' in publications, the (mathematics) research literature. If we could identify and acquire creative 'super-smart techniques' we could think of hiding them away in the super-structure that makes use of component AI systems in our 'naïve architecture'. This is a 'big if' where 'voices in heads' are concerned. Could the appearance of a 'first SI seed' from, say, a genius-trick, something like a 'Poincaré/Villani method', and/or some of the 'shaking up' manipulations we looked at, and/or some random fluctuations or mistakes, and/or even some trial and error moves, give a small advance beyond human level capability? And if we don't want that to happen, can we just make sure we reserve that 'super-structure role' for humans? These are good questions!

Some further insights into creativity have been obtained by addressing another question: how does the 'brawn' in our skulls support creativity? Neuroscience has not produced as much insight as one might have expected by now. But some possible directions for exploration can be found in, for example, the work of Arne

SUPERINTELLIGENCE AND WORLD-VIEWS

Dietrich[18] who is Professor of Psychology at the American University of Beirut, Lebanon. He has shed some light on certain aspects of this whole area. Traditional thought has it that imaginative ideas come from some random, blind perturbation of a gene-like or meme-like configuration of idea components, followed by selective retention of strings of 'genes' that have a high probability of being of 'high value' – usually with respect to some fitness function – coupled with a high chance of persisting. This search pattern is used in 'Genetic Algorithms' in some computer design exercises, for example. I mentioned earlier that I used it in the 1980s – to try to optimise the way data records should be placed on storage devices of computing facilities to give good speed of access.

Dietrich proposes two frameworks; one is characterised by that genetics-inspired *generate and test*, and he calls the other *predictive representation*. A particularly interesting suggestion is that a 'prediction imperative' explicitly accounts for what is seen as partial sightedness in some creative activities. People like Einstein seem to have been conspicuously motivated to seek to fulfil some clear intentions and to possess some limited but exceptional foresight. In his 'riding along on the crest of a wave' thought experiment, Einstein seems to have run prospective simulations off-line that feature expected future states and transition possibilities between them. There are broad parallels in the thinking of others who have looked at creativity patterns. For example, as we've seen, Roger Penrose says[4]: "*I imagine that the putting up would be largely unconscious and the shooting down conscious...One needs an effective procedure for forming judgements, so that only those ideas with a reasonable chance of success will survive.*"

Dietrich believes that some abilities have been integrated into the 'brawn', and an individual can supplement these gifts from accumulated experience. He believes that the creators erect their own *bridges* or *scaffolds* to give 'bread board' sketches of ways ahead, laced with 'black boxes' where details are unknown or of secondary importance. These may indicate, for example, that some intermediate forms can be jumped over without losing anything. Predictive representations may involve higher cognitive functions such as strategic planning and estimating, but some functions can work 'on autopilot', and concern things that can't always be readily verbalised, such as a sports skill or some imagined experience.

We saw in Chapter 5 that Einstein[19] believed that in his creative thinking: *"The psychical entities which seem to serve as elements in thought are certain signs and more or less clear images which can be 'voluntarily' reproduced and combined"*. I take this to include 'internal movies' and envisaged mechanisms. The 'how' information needed for emulation (as opposed to mere simulation) is useful here, and it might well trigger built-in intuitions and a *sort of* sixth sense in the designer/experimenter. Geniuses seem to be able to erect those mental bridges and scaffolding to work towards things of beauty or soundness even if some intermediate steps are very unclear or even plainly wrong, as Koestler noted. Even more pedestrian innovators, by representing goals and expected futures, have experiences of intention and foresight that seem to beyond the capabilities of other species, and perhaps SIs.

The last laugh. As a further thought in this chapter on lacks, let's look at very briefly at the lack of *humour* in artefacts. Humour is defined in Wikipedia as the 'tendency of particular cognitive experiences to provoke laughter and provide amusement'. Prospective artificial humour developers must be able to assess and quantify these cognitive experiences and perhaps decide with what sort of behaviour to replace laughter and amusement.

Most philosophers have been less than positive in their assessments of humour. Christian leaders in bygone times warned against either excessive laughter or laughter generally, and in monasteries and Puritan households there probably weren't many jokes flying around. This is not altogether surprising as some humour is sometimes seen as expressing feelings of superiority over classes of people we despise or whose fate we would like to avoid, or example. However, unlike emotions, laughter does not involve the motivation to do anything and an alternative way of viewing humour is, giving a release of excess nervous energy. Here I emphasise a different way of looking at humour which is probably the dominant theory these days and was taken by Kant and Schopenhauer who were interested in this topic. It says that our sense of humour, built-in as are other senses, is stimulated when we come across something incongruous – something that violates normal meanings or expectations or mental patterns. Kant, in his *Critique of Judgement*, referred to in Chapter 5, refers to this as wit or 'play of thought' and as 'free play of sensations' that 'promotes the feeling of health'. Humour in this sense is triggered by ideas being stretched and violated in the mind.

Simulating human humour as a complement to emotional and other attributes is seen as being an important goal for man/machine interaction. Some would say that a sense of humour is a key linkage between many of our human higher level capabilities. For example I heard somewhere of a school for children who are high-flying intellectually and creatively which uses 'a sense of irony' as a major selection criterion for aspiring 10 years old entrants. It seems somehow to be linked to consciousness and also creativity, with a little bit of emotion and perhaps even some morality thrown in. Telling jokes is even seen by some as being a special sort of creativity. So this is something that gets to the heart of much of what we're talking about here, especially if we decide to make machines 'more human-like' as well as more human friendly.

As a step towards improving this, some researchers have set up computers that can identify jokes based on 'sound puns' – similar sounding words with different meanings. When a machine hears a word in a 'story' that does not fit with in the rest of the context, the program searches for similar sounding words, and if a similar sounding word makes better sense, it identifies the story as a joke. It then responds appropriately; by giggling uncontrollably. There has been research to help uncover ways to imitate humans' humour. For example, Igor Suslov suggested a 'Computer Model of a Sense of Humour' in the early 1990's[20], and he has developed his ideas since. In Suslov's scheme, the humorous effect is said to be a: *"...specific malfunction in the processing of information, conditioned by the necessity of a quick deletion from consciousness of a false version"*. He says that it has a biological, or in his terms, evolutionary, origin rather than being a product of civilisation, and he argues that it leads to a more effective use of brain resources. Humour certainly plays a social role, though, and it seems clear that, as Freud claimed, the pleasure obtained from laughing underpins a sense of humour. Suslov says that its role in speeding up the consciousness of assimilation of information and effective brain usage are important in this. It is a positive thing: *"..if laughter afforded displeasure the social function of humour would change: the society would try to get rid of it by censorship, prosecution of witty people and so on"*. For the best effects, among other things, timing, known to be an essential feature of a human comedian's presentation, is usually looked for in addition to that: *"deviation from the norm"*. So it is clearly an interesting and hard challenge to give machines an appreciation of our sense of humour. In the movie *Bicentenial Man*, in an attempt to teach the 'hero robot' what a joke is, it's asked: *"Why did the chicken cross the road?"* Its reply is: *"One does not know, sir, possibly a predator was behind the chicken, or possibly there was a female chicken on the other of the road, if it's a male chicken. Possibly a food source, or depending on the season it might be migrating. One hopes there's no traffic"*. The robot can't see why the funny response *is funny*, and it's hard to see how that sense of humour could be transmitted to it.

Spotting textual jokes based on puns is relatively easy to automate. Computer-generated jokes are an entirely different matter. Elsewhere[21] in his outputs in this area, Suslov discusses the possible materialisation of a joker based on Hopfield neural networks. However, although simple jokes based on puns involving single words can be produced by 'a computer with humour' based on existing programs for machine translation, the fact remains that: *"teaching a computer to react on the more complex samples of humour related with the higher levels of information processing...looks incredible at the present time"*. Suslov reckons that this would need a lot of advances, and again it is not entirely clear what Suslov envisages as the sequence of steps towards this. *"In order to do it, one should reveal a complete set of images the average human brain contains and establish the correct associative links between these images. This would require many years of work of psychologists and programmers"*.

Others have developed software which generates witty one-liners in the form of surprising comments to follow up other statements. They say though, that really funny joke software would, like Emily, need to have, or to acquire, cultural awareness, but short of 'letting the 'bots out' in the way we've seen earlier in the book, this will be hard to do.

Making good these 'lacks' and providing the capabilities envisaged for the machines would push progress. The desirability of arranging for mutual understanding between agents of different kinds is something that seems to be self-evident, and these sorts of lacks currently impede this, and they will probably do so for some time to come. Of course, making good the 'lacks' would not ensure that a robot or an artilect has a cosmos-view, but realising this functionality is arguably an important step towards the Singularity that can't be avoided.

References

1. Nicolelis, M. A. L., Mind in Motion, *Scientific American* 307, September 2012.
2. Brooks, R. I., Rodney Brookes, Am a Robot, *IEEE Spectrum*, Jun 2008.
3. De Garis, H., What if AI succeeds?, The Rise of the twenty-First Century Artilect, AI Magazine 10.2, 1989.
4. Penrose, R., *The Emperor's New Mind*, Oxford University Press, 1990.

SUPERINTELLIGENCE AND WORLD-VIEWS

5. Blackmore, S., *Conversations on Consciousness*, Oxford, 2009.
6. Bisson, T., Meat, http://www.terrybisson.com/page6/page6.html, (originally in Omni Magazine, 1990.
7. Camp, J., Decisions Are Emotional, not Logical: The Neuroscience behind Decision Making, *BIG THINK*, June 11, 2012. http://bigthink.com/experts-corner/decisions-are-emotional-not-logical-the-neuroscience-behind-decision-making
8. Damasio. H., Damasio, A. R., Emotion, Decision Making and the Orbitofrontal Cortex, *Cereb. Cortex*, 10 (3), 2000.
9. Koenigs, M., Young, L., Adolphs, R., Trane, l. D., Cushman, F., Hauser, M., Damasio, A., Damage to the prefrontal cortex increases utilitarian moral judgements, *Nature* 446, 2007.
10. Aleksander, I., *How to Build a Mind: Toward Machines with Imagination*, Columbia University Press, New York, 2001.
11. Bagai, V., Shanbhogue, V., Zytkow, J., Chou, S., Automatic theorem generation in plane geometry, In *LNAI 689, Learning And Adaptive Systems I, Methodologies for Intelligent Systems*, eds Komorowski, J., Raś, Z. W., Verlag, S., 1993.
12. Colton, S., Computational Discovery in Pure Mathematics, In *Communicable Scientific Discovery*, eds Dzeroski, S. and Todorowski, L., Springer LNAI 4660, 2007.
13. Haase, K., Discovery Systems: From AM to CYRANO,*MIT Artificial Intelligence Laboratory, Working Paper 293*, March 1987.
14. Turing, A., Intelligent Machinery, *NPL report 1948* 1-20 HMSO, 1948.
15. Bell, J. S., Speakable and unspeakable in quantum mechanics, in *Speakable and Unspeakable in Quantum Mechanics-Collected Papers on Quantum Philosophy*, Cambridge University Press, 1987.
16. Poincaré, H., *The Foundations of Science*, first published in Paris in 1908, translated from the French by Halstead, G. B.
17. Villani, C., *Birth of a Theorem: A Mathematical Adventure'* Trans M de Bevoise, Farrar, Strauss, Giroux, 2015
18. Dietrich, A., Haider, H., Human creativity, evolutionary algorithms, and predictive representations: The mechanics of thought trials, *Psychonomic Bulletin & Review*, 22: 4, 2014.
19. Hadamard, J., *The Psychology of Invention in the Mathematical Field*, Princeton University Press, 1945.
20. Suslov, I. M., Computer Model of a Sense of Humour, (parts I and II). *Biophysics* 37, 1992.
21. Suslov, I.M., Can a Computer Laugh? (English translation from Russian available.), *Computer Chronicle (Moscow)* issue 1, 1994.

Chapter 14

What do *You* Think?

When the machine called Ava in the movie, *Ex Machina*, escaped and started its observations at the traffic intersection, it could have looked[1]: *"...to the multi-faceted and irregular results of observations for ....suggestions of overall structures and significant generalisations..."*. However, to become an SI under Bostrum's definition it would seem to be better if it did not waste time in 'grinding out' trivia such as often appears in studies for The Ignobel Prizes, organized by the magazine *Annals of Improbable Research*, which are for 'achievements that make people laugh, and then think'. J. Trinkaus, of the City University of New York, was a recipient of one of the prizes, and he has produced reports on his research on singularly uninteresting aspects of a huge diversity of ostensibly divisive and odd topics, such as: Brussels sprouts, wrapping-tissues and Stop Sign compliance. For one study on baseball caps[2] he found the energy and interest to watch: *"...407 people wearing baseball-type caps...in the downtown area and on two college campuses (one in an inner borough and one in an outer borough) of a large city"*. He came up with the different percentages of subjects who wore their caps back to front at the different sites. He has also obtained insights such as the percentage of drivers who don't bother to clear the snow off their car roofs and the percentage of commuters carrying lockable, as opposed to lockless, briefcases. Ava could easily conduct studies like those. Perhaps she/it could establish the percentage of drivers who do not stop completely at a particular stop sign, or confirm a previous finding – that women in vans are often the least law abiding of drivers. Without a cosmos-view to direct her/its attention, Ava might be interested in a fact like 'green cars tend to have licence plate numbers where the first digit is an even number', or it might even notice that men walk faster than women.

SUPERINTELLIGENCE AND WORLD-VIEWS

Genius level results would be unlikely to come from this, though. Having world-views would help Ava or any other SI hopeful to be more selective in choosing what sort of patterns would be rewarding, in accordance with her/its purposes and her/its 'something more important', if any. This illustrates one of the problems that automated knowledge discovery in general faces, for example in scientific discovery from large data repositories. The hard part is not so much finding the 'needle in the haystack' – although this may take up lots of research resources. A harder problem, that I tried to introduce in Chapters 4 and 5, is to find the right haystack on which to concentrate the search, and to obtain 'a magnet' to help in the search. Having a powerful magnet would be useful, and 'pointing that magnet at the right sized haystack' is important[3]. Thousands of patterns or 'items of knowledge' can be gleaned from repositories using current data mining programs, but how do we focus on what's 'interesting' and significant? Statistical measures of significance and definitions of interestingness are of help here, but more subjective measures derived from the analyst's domain knowledge, beliefs, values and expections are unavoidable.

I set out with two very ambitious intentions in Chapter 1. My first task specification was concerned with exploring the possibility that we will soon be able to design and build very smart mechanical agents, the SIs. There has been impressive progress in AI, if only within the scope of BAAs. However, SI capability, complete with genius-tricks, is still quite a bit away. There's also the question of whether the machines could self-reproduce and find some way of increasing their population and, for example, find an Ingredient x to live for, that I side-stepped in Chapter 1. Getting W*s to self-replicate is impossible FAPP as yet. Admittedly, there are, for example, robots made up of identical cubes called *molecubes*, each having the same program, which permits replication[4], and there is a commercially available 3D plastic-printing machine which can replicate most of its own components. And for decades theoretical results have existed on self-replication, self-organisation and self-assembly, such as von Neumann's[5] analysis. But we're clearly very far away from the sort of artefact reproduction that would be needed for W*s. My second task specification was complementary – about looking at how the SI concept impacts on our own human world-views. There is a big question around how 'standard' narratives that we use, say those concerning science and spirituality, would be impacted if the dream, or nightmare, of SIs were to become a reality.

A look at the question of whether a machine would ever *need* to have world-views sheds light on possible approaches to both these issues. I say: 'Yes – if we're going to reach machine geniuses and the Singularity; No – if we're using machines only as tools or pets'. Overall, my main conclusion is that there will not be a Singularity, for good technological reasons, but also very largely due to the catastrophic effect the existence of SIs would have on our epistemology – especially in traditional faith stances. So they won't need world-views.

Is the Singularity imminent, or even likely at all? Let's look more closely at that first task I set myself. The objective time-space-matter cosmos-views we looked at assert, per Deutsch, that nothing smarter than humans is possible, and as Tipler/Teilhard/etc also claim, the future lies with transhumans – not SIs. Another negative 'pull' comes from the fact that, as I argued in Chapters 4 and 5, geniuses such as Einstein, Picasso, et al, needed or need world-views. Where would SIs source appropriate world-views? Plenty of questions are left unanswered by the two options we have: an 'Ava model' of seeding and exploration could be used, or human designers could give the SIs 'fixed world-views', in which case I've argued that the term SI might well be a mis-nomer.

Thinking about possible spiritual world-views for SIs leads to further problems. Would SIs *worship* anything? And what would the spiritual status of SIs be? Why would they ever need a concept of sin and salvation in their SI world? For example, why would a very advanced Ava or Sonny, or even a W*22, ever need to be *regenerated* in the way, say, evangelical Christians[6] talk about? There was no original sin for this 'species'. The equivalents for machines of other related theological concepts such as illumination, revelation and assurance that are said to be available for humans, and which arguably represent a definite 'epistemic plus' for Christian believers, are hard to imagine.

Moreover, Christian believers would have to change their theological world-views as 'eternal verities' would have to be revised, and that alone rules out effectively, for them, even consideration of the accommodation of SIs. Eternal verities by their very nature are not things that are discarded easily. Those world-views point away from aiming to make the 'better than human' agents that the Technological Singularity would usher in, as some say that we can't better humans as the pinnacle of the created universe. Any attempt at this would, by definition, fall short because it fails to incorporate many things of importance, such as the purposes and plans of God, and the transmission of his image to humankind. Then again, would some humans be induced to worship SIs if they existed? It is very easy for humans to be led astray into false ideas about God, and artefacts have outstanding ability to encourage that. There would also be an increased possibility that we could start to think of God in the image of man, albeit as a very advanced one, which is warned against in the Bible. We could only base our theology on SIs if we neglected 'illumination, revelation and assurance', and perhaps tried to arrange the future à la Tipler. This means that the epistemological problems that SIs would raise loom large – especially with respect to those theological narratives that would be hard to shift from the psyche of the believers.

A more earth-bound negative for the goal of, say,' SIs by 2018' is that technology will not be good enough. We've thought of the difficulty of finding genius-tricks for robots, even if we did find a way of discovering more fully what made people like Einstein and Picasso tick and how they saw the world, and I also sketched out a selection of specific 'lacks' in current AI, ranging from emotions through creativity to humour. These stand out alongside the 'haystack identification problem' and the very hard 'self-replication problem'.

My conjecture is that the gap between current mechanical agents and what is needed for machines to be 'like us' or 'better than us' points to the conclusion that humans and machines must always remain different categories of agent, given the theological and technological gaps. This conjecture has not been proved – but it is as well-supported as many other 'effective impossibilities' we humans come across from time to time, and manage to live with satisfactorily. For our conjecture there is a mass of evidence to support the conclusion that the problem involved seem to be *just as intractable* FAPP as many other problems that are believed to be effectively intractable. There is very clear evidence to support a belief that mankind has characteristics, especially in the spiritual realm, that are very special, and that they can't be understood fully, much less reproduced for our artilects.

I have to point out, however, that I have learned to be cautious about saying 'never' in matters of research and development in technology! The history of mankind's journey and the build-up of knowledge therein is strewn with 'can't do's' that have been overturned later. The Simon Newcomb Award is 'presented' for published arguments against AI that appear very silly to the judges, and to get the award they say, very scathingly, that these should demonstrate that: *"the writer's confidence in his views seems to arise from his ignorance of the subject"* [7]. In 1888, Newcomb, a highly respected scientist, stuck his neck out by saying that nearly all that could be known about astronomy was already known, and that statement alone is probably enough to make researchers cautious in making negative science and technology predictions. But the award is named after Newcomb because of another misjudgement claiming that manned flight was impossible because even you could get off the ground and move around, you couldn't stop. *"Once he slackens his speed, down he begins to fall. Once he stops, he falls as a dead mass"* [8]. A desire to avoid having dubious awards named after them makes would-be technology prophets in particular reticent about making 'can't do' claims!

However, all of the theological difficulties alongside those, at least very, very difficult, technological hurdles, combine to weigh heavily against SIs.

The impact of SIs on Human Cosmos-views? Is there any more general fall-out for our human cosmos-views? Our discussions have certainly brought our own world-views into the spotlight. We humans are on a journey as a species, so are our human cosmos-views affected by considerations of the challenges and possibilities related to SIs? Is there anything we can learn or attend to in the realms of knowledge, belief, prediction and explanation? What about our values and goals? This demands attention on our part, perhaps using Vidal's framework referred to in Chapter 3.

First of all, it is important to acknowledge the fact that it is manifestly possible for humans to be *brainwashed*, maybe using subtle or forthright psychological and other techniques to subvert a subject's sense of control over their own thoughts. Subjects are persuaded to conform to the world-views of manipulators. Conversion to strange religious sects is said to be, in extreme cases, by compromising a subject's freedom of choice and action. Modification or distortion of the subject's behaviour, beliefs, goals and values, may be sought. Aspects of a subject's neurological make-up can be manipulated, or the covert exploitation of the unconscious can underlie this. Brainwashing is probably too strong a word for the more familiar and friendly influences and persuasions that there are on the world-views of all human beings, but everyone should be interested in whether or not they have accepted 'given' world-views too readily. Conditioning certainly occurs where children are drip-fed from infancy, by media, school and even parents. We can be negatively anchored in all aspects of our world-views, and in particular in the scientific and spiritual narratives we're given.

Let's take an example from science. These days it is often, by default, taught that all development needed to get to, for example,

SUPERINTELLIGENCE AND WORLD-VIEWS

our human level of sentience from 'dust' is explained as step-by-step random mutation and selection by the environment. The remarkable flow and skew towards life, contrary to chance and chaos, that was required to even get early life from molecules, is simply ignored. Without the concept being taught of mind and intelligence being stitched into the universe's 'raw material', the learners' knowledge will be deficient. Unbalanced, closed indoctrination in general tends to distort conceptions. Openness is called for. Similarly there should be no lack of openness in, say, religious senses of truth. There are, of course, plenty of examples of over-indoctrination in religions. Since spiritual considerations are often taken to be of primary importance, a person might be expected to swallow what they are told without question or critical examination. This is often transmitted in a way that can work mischief on naïve minds. It is therefore essential that received views be subjected to testing against reason, general experience, and the given revelation, but also against experimental evidence, although the fact that those two prominent narratives remain beyond any proof methods currently available in the human condition appears to be axiomatic.

The consensus of the 'best informed' people these days – those open to all the evidence – is that science is pointing to a finite universe of incredible complexity, and thus it demands extra-material explanation. Leading physicists have placed importance on the fundamental role of information. The nature of the mind behind the universe rather than its existence still gives space for important differences in belief. For example it fits relatively well with the accounts of primordial workings of 'wisdom' in the Jewish faith and *logos* or 'the Word' in the Christian faith. However in the latter case the belief is that the Word 'became flesh' to exclusively secure that salvation referred to in Chapters 1 and 8 above. Natural philosophy and at least some traditions of theology have a nuanced relationship that is worthwhile to nurture and if possible develop, but at times it gets strained.

A second way that we can be attentive to our world-views and cosmos-views is to be open to the old as well as the new as we saw in Chapter 6. It was natural for the beliefs, values, norms, and institutions of a number of societies to alter in some important ways as their social, technological, economic, political, and environmental factors changed. Values in particular have been heavily impacted over the epochs. For example, the basis of early 'solutions' to the Big Question of giving an explanation of the world, especially explaining those aspects that influenced day to day living FAPP, tended to be animistic. The belief was that spirits inhabit virtually everything in the world of nature. Values were set accordingly, maybe being specified by a misled or misleading shaman, but as time went on 'the received wisdom' was given a decreasingly easy ride. Explanations were increasingly challenged partly due to the improvement of information storage and dissemination methods. However, what has never disappeared is the fact that humans have aspirations, appetites and ambitions, kinds of essential needs and, perhaps, 'somewhat more luxurious' desires, that are, on the face of it, irrelevant to our survival as individuals or as a species. First-order, mundane, non-reflective, animal knowledge-level, struggles must be addressed, but there's more to human life than that, as we saw in Chapters 6 and 7. So it is important to acknowledge, evaluate and learn what we can from what we know of, for example, earlier times as we touched upon in Chapter 10. The best societies weren't completely open, but neither were they completely closed or static.

Looking to the future it is interesting to consider if and how the patterns we've seen in technology and its assimilation into, and consequences for, society are likely to change. This is particularly so with respect to the development of AI. In Isaac Newton's times there would have been technology, artefacts and understanding – boats, cathedrals, scientific laws – that would have greatly astounded any 'thought experiment' time-travelling hunter gatherers from early BC. Progress from the discovery of fire to the discovery of the law of gravity was, however, pedestrian compared to that between Newton's time and our own. The 'astounding' would have been much greater, if a time-traveller from the 17th century could speak to someone here who had been in Australia yesterday, and play Go against a shiny, flattish, highly portable rectangle. Seeing footage of action from the First World War and looking up Google to find how to get to a hotel would presumably seem miraculous to them. The ideas of quantum mechanics, nuclear power generation and space stations would blow their minds. A *sort of* 'law of exponential returns' has also meant that the rate of progress in knowledge generation for the 20th century was several times the average rate.

Typical modern humans are, therefore, 'well informed' compared to even geniuses from the past, like Newton. But many are disinterested in the details of the very abstract reasoning needed for, say, Andrew Wile's proof of Fermat's Last Theorem, or quantum mechanics, or NP completeness theory. They are, of course, interested in the fruits that such understandings can bring in their lives, and they know the rough pattern of how such knowledge is accumulated, and they do have awareness of STEEP factors in their lives.

Collectively mankind understands all of this and SIs would just add to the man-made world. We still work with very limited information, however, and questions of origins, purposes, morality, and so on are opaque to the scientific method.

So there is much in life that I can reach but not fully grasp. I often use a statement by Henri Blocher about an enigmatic issue[9]: *"We should not be embarrassed to conclude with uncertainty: it is a mark of mature faith, properly based on adequate evidence and serenely bearing the tensions of the pilgrim's progress by faith, not sight"*. Some 'outside the reach' knowledge is obtained in an extra-scientific way – through direct disclosure by a higher authority. Yet it is not simply filling 'the gaps' – intellectual difficulties that are left when science has done its impressive job of explaining what people observe. The testimony of those who have experienced personal assurance is one evidential input to this. They claim that the extra dimension offered by enlightened faith enriches a person's existence immeasurably.

In conclusion, it is still interesting to consider the question: if, against those odds, SIs were to appear, what would we use those conscious, super-smart artilects for? Or what would they use us for? And how would they be controlled? These are hard, but, thankfully, probably hypothetical questions. However there is nothing to stop researchers with specialisms like mine being challenged in a positive way by such questions and using the 'missing capabilities' list as conceptual stretch targets when pushing forward in continuing to design and make tools that can improve mankind's lot.

So I would like to address a final question that is close to my own heart. What should we researchers in AI work at? Was my life in the lab a waste of time? Like nearly all the computer scientists I know, I have been working, in my own research, towards the goal of 'machines as tools'. The difficulties of materialising SIs and/or organising 'our next evolutionary step' towards the Omega Point mean it always was for me, and it will remain, at best, an unreachable target. Focus on the spectacular is often the norm in research, but few people would object strongly if research were to be heavily skewed towards more mundane struggles than the development of SI. AI techniques are used routinely to improve efficiency and effectiveness in applications in (eg) medicine, education, transportation, food-production and other relatively mundane applications. Pollution, hunger, disease and many other ills can be tackled today in ways that would be inconceivable 100 years ago, and the old Jewish concept of 'world repair' in its broadest sense would be an admirable goal for AI.

It seems to be axiomatic that the quality of the physical and spiritual aspects of the lives of humans should be prioritised. The satisfying of widespread needs should be put well before the optimizing of idiosyncratic desires. Persistent lightness of foot will be needed if this is to be done, though, as challenges and opportunities for positive changes of direction and paradigm are never likely to disappear fully or be fully controlled. Creativity is an essential in human designers, and it can be energized by technology, even if it is never actually supplemented by mechanical geniuses! This gives a more positive picture as we go forward than that of being increasingly tied up by and possibly coerced by artefacts. Attainment of all of these possibilities is in considerable doubt, though, when we consider history and experience, where greed and vanity are seen to be great influences.

SIs and world-views touch on many wide-ranging considerations, and at the deepest level, we're left wanting more than we're given on any individual concept. However my hope is that some accumulating light can be shone on the various perplexities I've aired, and that possible gaps in human world-views have been spotlighted.

References

1. Goodman, N., *Ways Of Worldmaking*, Hackett Publishing Company, Inc, 1978.
2. Trinkaus, J, Wearing Baseball-Type Caps: An Informal Look, *Psychological Reports*, 74:2, April 1994.
3. https://blogs.rsa.com/stop-climbing-through-the-haystack-to-find-the-needle-use-a-magnet/
4. Zykov, V., Mytilinaios, E., Adams, B., Lipson, H., Robotics: Self-reproducing machines, Brief Communications, *Nature*, 435, May 2005.
5. Sayama, H., Von Neumann's Machine in the Shell: Enhancing the Robustness of Self-Replication Processes, in *Artificial Life VIII*, eds Standish, Abbass, Beda, MIT Press, 2002.
6. Nicholls, S. J., *An Absolute Sort of Certainty*, P and R Publishing Co, New Jersey, 2003.
7. Hayes, P., Ford, K., The Simon Newcomb Awards *AI Magazine* Volume 16 Number 1,1995.
8. Newcomb, S., The Outlook for the Flying Machine, *The Independent*,1901.
9. Blocher, H., The Theology of the Fall, in *Darwin Creation and the Fall*, eds Berry, R. J. and Noble, T. A., Apollos, 2009.

Index

A-schema 34, 37, 43, 137, 146, 147, 156, 160
Adam 112ff
Agent (eg) definition 4; basic artificial (BAA) 4 (also 22-3, 43, 193); artificial 1; better than human 194; human 3; intelligent 6, 94, 157, 176; supernatural 4-5
AI 16 (McCarthy definition), 6, 9, 17, 19, 24, 120, 127, 140, 143, 171-2, 174ff, 193, 195, 198ff
Allen Woody 95
AlphaGo 20, 22, 24, 120, 178
AM 184
Annals of Improbable Research 192
Apocalypticism Ch 13, p 174–
Apostel L 33, 146, 156, 160
Armstrong K 112-3
artificial life (see *self-replication*)
artilect (definition) 43-4
artificial intelligence (seeAI)
Austin Steve 92
Ava 3, 22, 25ff, 137, 139, 147, 151, 179, 192ff

BAA (see agent)
Bach, JS 8, 24, 27, 30, 172
Bacon F 54, 166
Barth K 115
Bell JS 88, 184
Bender 117ff

Berlin Wisdom Paradigm 128
Bisson T 179
Blackmore S 157
Blocher H 114ff, 199
Bloom A 79ff, 86, 91, 97, 114, 121, 147
Bostrum N 6, 7, 44, 66, 91, 178, 192
Brassai G 57, 58, 68ff
Bronowski J 32, 84
Brooks R 175ff

Carey S 170
Cepheus 5, 8, 19ff, 43, 183
Chesterton GK 40
Christianity, Christ, Christian 12, 51ff, 87, 92ff, 99ff, 114ff, 150, 174, 188, 194, 197
Cobbett W 46-7, 50, 129, 140, 147, 149
coherence 49, 53
consistency inconsistency 23, 42-3, 53, 72, 74, 81, 168, 171
Cope D 30, 74-5
cosmos-view (see A-schema) 40ff, 46ff, 60, 71, 73, 79, 81ff, 86ff, 95ff, 101, 103, 105, 108, 111, 113ff, 122, 124, 125, 128, 130ff, 140, 144, 147ff, 160, 166, 176, 190, 192, 194, 196ff
cottager 47ff, 143
Cyrano 184

da Vinci L 59, 166
Damasio H 180
Darwin C 8
Das Nichtige 115
data mining 163, 193
Dawkins R 157
de Chardin T 40, 87; see also Teilhard 90, 102–3, 132, 176, 194
de Garis H 43
Deep Blue 2, 5, 8, 19, 22–23, 29, 40, 116, 146, 178
deism 104
del Sarto A 10
Deutsch D 13–14, 39–40, 69ff, 83–84, 86ff, 95, 102, 105, 119–20, 125, 142, 161, 175, 178, 194
dialetheism 169
Dietrich A 187
Dilthey W 57
Dyson F 100

Easter Island 83–86
Eco U 98, 101, 165
Einstein A xi, 5, 8, 23–4, 27, 44, 55ff, 64–5, 68ff, 79, 83, 88, 99, 166, 182, 184ff, 194–5
Ellul J 143
EMI 30–31
Emily 2, 5, 8, 19, 30–31, 43, 68, 70, 74ff, 76, 116, 143, 183, 190
epistemology 34, 90, 132, 134, 137, 147, 194
ESA 63–64, 119, 131, 170
evolution/evolutionary 39, 87, 104, 109–10, 119, 132–3, 138, 157–8, 176, 189, 199
Ex Machina 2, 22, 116, 138, 179, 192

Fall (of Mankind) 109, 113ff, 119
FAPP ('for all practical purposes') ix, 49–50, 57, 80, 82, 98, 101, 127, 136, 148, 154, 160, 164, 170–71, 193, 195, 197
Flojo 152
football/footballer 24, 28, 64, 86, 105, 172
flops 177–178
Freire P 126
Freud S 9, 41, 80, 136, 189
Futurama 117–8, 120

Galileo Galilei 53ff, 68, 83
Gee H 108, 111, 133, 161
global warming 3, 25, 43
Gödel K 166, 168
Goodman N 42, 67, 71
Greenfield S 38–9, 94

Hadamard J 69
Hawking S 6
Heaney S 48, 50, 56, 143
Heisenberg W 99
Hokusai K 72–3
human development 111, 156
human genome 110

I Robot 2, 26, 33, 116
Ignobel Prizes 192
Ingredient x 27–28, 40, 99, 193
innateness 152

Jaspers K 85
Jeans J 99
Jennings K 20
Jesus (see also Christianity) 12, 87, 100, 104, 116
Johansson D 109

Kahneman D 155
Kant I 67–8, 188
Kasparov G 19
King Solomon 95, 121–2
Koestler A 51–3, 103, 147, 149, 184, 188

Lakof G 154
Leakey M 109

Lightman B 97
Linpack Scale 141
Livingstone D 50ff, 104, 129, 147, 149–50
Lloyd S 11–12, 14, 17, 89–90
Lorenz K 156
Lucy 109ff

mathematics/maths 1, 8, 12, 21, 25, 53, 55, 66, 70ff, 90, 98, 99, 153, 155, 160, 162ff, 178, 184ff
McCarthy J 16
Michelangelo 70
Midgley M 132–3
Miller H 57
Moravec H 176
Morgan C 115–6
Mozart W 5, 30, 59, 75
muddy (problems) 24, 126
music/musical 1, 19, 30–31, 57, 67, 70–71, 73ff, 100, 183

natural language 16, 20–21
Newcomb S 195–6
Newman J 141, 163, 166
Newton I 24, 27, 59, 64–5, 75, 153, 162, 166, 184, 198
Nietzsche F 28
Nobel Prize 88
Noether E 55–56
neural correlates 180
neural substrates 129, 157

Omega Point Theory Cosmologies x, 13–14, 40, 86ff, 102, 132, 142, 199

Pannenberg W 40, 102–3
Penrose R 53, 59, 76, 187
Philips W 169
philosophy 10, 25, 28, 57, 101, 122, 131, 134–7, 141, 145, 149, 197
Picasso P xi, 8, 57ff, 65ff, 79, 83, 166, 182, 186, 194–5

Pinker S 11, 17, 59, 90
Poincaré H 85–6
Planck M 99–100, 134
Polanyi M 11
Pope A 166

quantum computer 11, 89
quantum mechanics 72, 88, 100, 164, 198

Rano Raraku 84
Rapa Nui 83–4
realism 42, 107, 135
Redi F 162
relativism 79, 81–2, 129, 169
religion 1, 25, 27, 32, 38, 43, 52, 54, 57, 84, 95–7, 101–2, 104, 118, 132, 197
Rokeach M 148–9
Rollin Justin 3, 8

Sacks J 86, 104–5, 112
Samuel A 22
Schaeffer F 54, 65
science fiction 6, 107, 113, 118
science historical 108, (110–12), 133
science operational 108, 110–12, 133
scientific method 132–134, 199
Sedol L 20
self-replication 193, 195
Shaw G 56, 97
SI (see superintelligence)
Singularity 2, 24, 31, 120, 172, 175–6, 179, 190, 194
Smith A 147–8, 152
Socinius 100, 104
Socrates 121
Solipsism 34
Solomon (see King Solomon)
Sonny 2, 25–27, 29, 33, 147, 194

Sosa E 78
spontaneous generation 99, 162
Stace W 96
Standard Model general 82; physics 12, 87; human development 108, 110–12, 163; sin 114, 116; cosmology 114, 116, 163–4
Star Trek 9, 118, 120
Stradel Cave 9
superintelligence, superintelligent ix, 6 (definition) 11, 18, 24, 31, 44, 60, 105, 177–8, 182, 184
supernatural xi, 4, 11, 14, 23, 43, 94–5, 99, 103–4, 135
Suslov I 189ff
synthetic biology 3, 17

Teilhard *see de Chardin*
Tertullian 54
theism 103
theology, theological ix, xi, 11–13, 25, 32, 40, 46, 50, 54, 78, 87, 90, 94, 96, 100, 102, 107ff, 112ff, 174, 176, 194ff
Tianhe 141, 177–8
Tipler F 12ff, 40, 52, 87, 89ff, 102, 105, 113, 119–20, 132, 175, 176, 178, 194–5
transcendence, transcendental xi, 11, 13–14, 86, 103, 105, 107, 109, 122, 132, 134, 139, 142
transhumans 90ff, 175–6, 194
Trinkaus J 192
Turing A 27, 29, 151, 184; Award 12; Test 179; machine 89, 99
Tversky A 155
Tyndall J 97

values ix, x, 1, 3, 8, 24, 26–7, 29, 33ff, 41, 43, 46–7, 50, 53, 55, 58, 66, 71, 73, 80, 82, 85, 96–7, 99, 114, 116, 126–7, 129, 135–6, 138–9, 142, 144, 146ff, 157–8, 168, 177, 179, 181, 193, 196–8
van Helmont J 162
V'ger 2, 118, 120
Vidal C 33–4, 42–4, 136–7, 196
Villani C 71, 186
virtual reality 7, 75, 88, 91, 142, 175, 178
von Goethe W 103
von Neumann J 22, 193

Ward K 10, 97
W^* 24, 146–7, 150–52, 182, 193
Watson 2, 5, 8, 19–21, 23–4, 27, 29, 36, 37, 76, 116, 118, 120, 141, 146, 152, 178, 183
Wheeler JA 65, 88
Wiles A 166–7
wisdom 6, 7, 44, 80, 121ff, 137–8, 149, 167, 176, 197–8
Wittgenstein L 86
world-views (see A-schema) ix-xi, 1, 3ff, 7–8, 9 (Freud definition), 10–11, 14, 16, 18, 23ff, 31ff, 40ff, 49ff, 57ff, 66–7, 72, 75–6, 78, 85, 89, 95ff, 103, 105, 108, 111–2, 117–118, 121ff, 136 (Vidal definition), 137, 139, 140, 141, 146ff, 151, 157–8, 168, 171, 175, 177, 193–4, 196–7, 200